CHEMICAL NOMENCLATURE USAGE

ELLIS HORWOOD SERIES IN CHEMICAL SCIENCE

KINETICS AND MECHANISMS OF POLYMERIZATION REACTIONS: Applications and Physicochemical Systematics
P. E. M. ALLEN, University of Adelaide, Australia and
C. R. PATRICK, University of Birmingham
ELECTRON SPIN RESONANCE
N. ATHERTON, University of Sheffield
METAL IONS IN SOLUTION
J. BURGESS, University of Leicester
ORGANOMETALLIC CHEMISTRY – A Guide to Structure and Reactivity
D. J. CARDIN and R. J. NORTON, Trinity College, University of Dublin
STRUCTURES AND APPROXIMATIONS FOR ELECTRONS IN MOLECULES
D. B. COOK, University of Sheffield
LIQUID CRYSTALS AND PLASTIC CRYSTALS
Volume I, Preparation, Constitution and Applications
Volume II, Physico-Chemical Properties and Methods of Investigation
Edited by W. GRAY, University of Hull and
A. WINSOR, Shell Research Ltd.
POLYMERS AND THEIR PROPERTIES: A Treatise on Physical Principles and Structure
J. W. S. HEARLE, University of Manchester Institute of Science and Technology
BIOCHEMISTRY OF ALCOHOL AND ALCOHOLISM
L. J. KRICKA, University of Birmingham and
P. M. S. CLARK, Queen Elizabeth Medical Centre, Birmingham
MEDICAL CONSEQUENCES OF ALCOHOL ABUSE
P. M. S. CLARK, Queen Elizabeth Medical Centre, Birmingham
L. J. KRICKA, University of Birmingham
STRUCTURE AND BONDING OF SOLID STATE CHEMISTRY
M. F. C. LADD, University of Surrey
METAL AND METALLOID AMIDES: Synthesis, Structure, and Physical and Chemical Properties
M. F. LAPPERT, University of Sussex
A. R. SANGER, Alberta Research Council, Canada
R. C. SRIVASTAVA, University of Lucknow, India
P. P. POWER, Stanford University, California
BIOSYNTHESIS OF NATURAL PRODUCTS
P. MANITTO, University of Milan
ADSORPTION
J. OSCIK, Head of Institute of Chemistry, Marie Curie Sladowska, Poland
CHEMISTRY OF INTERFACES
M. J. JAYCOCK, Loughborough University of Technology and
G. D. PARFITT, Tioxide International Limited
METALS IN BIOLOGICAL SYSTEMS: Function and Mechanism
R. H. PRINCE, University Chemical Laboratories, Cambridge
APPLIED ELECTROCHEMISTRY: Electrolytic Production Processes
A. SCHMIDT
CHLOROFLUOROCARBONS IN THE ENVIRONMENT
Edited by T. M. SUGDEN, C.B.E., F.R.S., Master of Trinity Hall, Cambridge and
T. F. WEST, former Editor-in-Chief, Society of Chemical Industry
HANDBOOK OF ENZYME BIOTECHNOLOGY
Edited by A. WISEMAN, Department of Biochemistry, University of Surrey
INTRODUCTORY MEDICINAL CHEMISTRY
J. B. TAYLOR, Director of Pharmaceutical Research, Roussel UCLAF, Paris and
P. D. KENNEWELL, Deputy Head of Chemical Research and Head of Orientative Research, Roussel Laboratories Ltd., Swindon, Wiltshire

CHEMICAL NOMENCLATURE USAGE

Editors:

R. LEES, C.Chem., M.R.S.C., F.R.S.H., A.I.F.S.T.

and

A. F. SMITH, C.Chem., M.R.S.C.

both of The Laboratory of the Government Chemist
London

Published by
ELLIS HORWOOD LIMITED
Publishers · Chichester

For the
LABORATORY OF THE
GOVERNMENT CHEMIST
London

First published in 1983 by

ELLIS HORWOOD LIMITED
Market Cross House, Cooper Street, Chichester, West Sussex, PO19 1EB, England

The publisher's colophon is reproduced from James Gillison's drawing of the ancient Market Cross, Chichester.

Distributors:

Australia, New Zealand, South-east Asia:
Jacaranda-Wiley Ltd., Jacaranda Press,
JOHN WILEY & SONS INC.,
G.P.O. Box 859, Brisbane, Queensland 40001, Australia

Canada:
JOHN WILEY & SONS CANADA LIMITED
22 Worcester Road, Rexdale, Ontario, Canada.

Europe, Africa:
JOHN WILEY & SONS LIMITED
Baffins Lane, Chichester, West Sussex, England.

North and South America and the rest of the world:
Halsted Press: a division of
JOHN WILEY & SONS
605 Third Avenue, New York, N.Y. 10016, U.S.A.

British Library Cataloguing in Publication Data
Chemical nomenclature usage. –
(Ellis Horwood series in chemical science)
1. Chemistry – Nomenclature
I. Lees, R. II. Smith, A. F.
540'.14 QD7

Library of Congress Card No. 83-77

ISBN 0-85312-368-3 (Ellis Horwood Ltd., Publishers)
ISBN 0-470-27447-6 (Halsted Press)

Typeset in Press Roman by Ellis Horwood Ltd.
Printed in Great Britain by R. J. Acford, Chichester.

Table of Contents

Introduction

The gulf that exists between experts on chemical nomenclature and day-to-day users of chemicals names has widened in recent years. The scientist with more than 60,000 common research chemicals at his command and the industrialist with more than 10,000 manufacturing chemicals finds increasing difficulty in keeping abreast of changes in systems of nomenclature even though these changes are designed to give a single unambiguous name.

The Laboratory of the Government Chemist in association with the British Crop Protection Council, British Pharmacopoeia Commission, the British Standards Institution, the Chemical Industries Association, the Chemical Notation Association (UK), the Pharmaceutical Society of Great Britain, the United Kingdom Chemical Information Service, and the Joint Nomenclature Panel of the Royal Society and Royal Society of Chemistry sponsored a Symposium designed to help overcome the confusion and misunderstandings that exist. The aim was to bring together users and developers of chemical nomenclature in an attempt to increase understanding of the subject and develop a common approach. The chapters in this book are edited versions of papers given at that Symposium. They cover many of the issues and problems involved in the use of chemical names, indicate changes in present thinking on nomenclature, review available sources of assistance and name and structure management schemes, and consider likely future developments. The text has been prepared as a contribution to the literature on the usage of chemical names comprehensible whether or not the reader attended the event.

R. Lees
A. F. Smith
November 1982

Note. Summaries of the discussions which followed the various presentations are available for the cost of return postage charge from the Editors at the Laboratory of the Government Chemist, Cornwall House, Stamford Street, London SE1 9NQ.

The Symposium

The Symposium was held at Church House, Great Smith Street, London SW1, from 24 to 26 March 1981. The aims were:

to assemble representatives of the users of chemical nomenclature and conceivers and providers of guidance on nomenclature rules;

to provide users with information on available systems;

to provide IUPAC and other relevant organisations with feedback from users;

to provide information on future developments; and

to provide a forum for discussion on the use of chemical nomenclature.

The Contributors

Mr H. J. Chumas is Director of Tariff Questions in the Custom Union Service of the Commission of the European Communities in Brussels. He is responsible for the management of the Common Customs Tariff, of tariff quotas and duty suspensions, and for customs and tariff questions in connection with international agreements. Mr Chumas has been the negotiator for the European Community on GATT multilateral trade negotiations. Before moving to Brussels, Mr Chuman was head of management services for Her Majesty's Customs and Excise and was a member of the United Kingdom Board for the Simplification of Trade Procedures.

Dr H.B.F. Dixon is a lecturer in biochemistry at the University of Cambridge. He worked for a period at the Institute of Molecular Biology in Moscow. His research interests are concerned with corticotropin, melanotropin, the specific modification and removal of the N-terminal residue of proteins by transamination, phosphonates and arsonates as analogues of natural phosphates, and ligand binding. Dr Dixon has been a member of the Editorial Board of the *Biochemical Journal* and is currently secretary of the International Union of Biology – International Union of Pure and Applied Chemistry (IUB – IUPAC) Joint Commission on Biochemical Nomenclature and of the Nomenclature Committee of IUB.

Dr H. Egan joined the Laboratory of the Government Chemist (LGC) in 1943 specialising in the analytical chemistry of foods, drugs and agriculture. He was appointed Government Chemist – an office that can be traced back to 1842 – in 1970 and served in this capacity until his retirement in 1981. At the time of publication he is a member of the Council of the Analytical Division of the Royal Society of Chemistry, visiting professor and external examiner for a number of universities, Chairman of the Editorial Board for Analytical Methods of the International Agency for Research on Cancer; and is a former member of the IUPAC Interdivisional Committee on Nomenclature and Symbols, and has served on a number of joint Food and Agriculture Organisation/World Health Organization committees.

Mr E.W. Godly joined LGC in 1950 and his experience as an analytical chemist covers textiles, foodstuffs, pesticides, fuels and alcohol control. In 1971 he joined the nomenclature section of that Laboratory and later was appointed Head of a new Chemical Nomenclature Advisory Service (CNAS). Mr Godly is a United Kingdom representative on the Chemists' Committee of the international Customs Co-operation Council. He is currently an associate member of both the Organic and Inorganic Commissions on nomenclature of IUPAC.

Mr E. Hyde is Managing Director of Fraser Williams (Scientific Systems) Ltd operating in the area of computer services for scientific users. Following work on organic research he became Head of Research Data Services of the Pharmaceutical Division of ICI. It was during this period that he developed CROSSBOW, a computerised structure handling system for chemical substances.

Mr R.T. Kelly is Scientific Adviser to the Greater London Council (GLC). His early career was spent in the refractory and concrete industry. He joined the Scientific Branch of the GLC in 1963 as Head of the Building Sciences Group and was appointed Scientific Adviser in 1974. Mr Kelly has served on many committees including those of the British Standards Institution (BSI), Concrete Society, Building Research Establishment, and Plastics Institute. He is currently the GLC Assessor to the Agreement Board. His particular interest in nomenclature is the use of terms and names in an unambiguous way. This is particularly relevant in fire fighting, and the handling of toxic waste where hazardous situations may develop through the use of imprecise terminology.

Mr M. Kenward studied physics at Sussex University and then spent three years undertaking research on fusion energy with the United Kingdom Atomic Energy Authority. He joined the staff of the weekly journal *New Scientist* in 1969 and was later appointed Editor. His particular interest lies in the use of clear, understandable nomenclature for scientific articles that are intended to be read by a wide cross-section of the public.

Mr C.D. Kopkin graduated in chemistry at London University prior to joining the Royal Air Force. On return to civilian life he spent four years in the chemical industry before joining the Patent Office in 1957. Mr Kopkin has worked on the examination of inventions in various fields of chemistry both for the United Kingdom and for the European Patent Office. He has devised a number of classification systems for use in various chemical areas concerned with patents. This work has involved co-operation with other national Patent Offices on the use of shared schemes.

Dr K. L. Loening is Director of Nomenclature of the Chemical Abstracts Service (CAS) of the American Chemical Society (ACS). He is Chairman of the

ACS Committee on Nomenclature and Chairman of the IUPAC Interdivisional Committee on Nomenclature and Symbols (IDCNS). Dr Loening is also a titular member of the IUB – IUPAC Joint Commission on Biochemical Nomenclature, an associate member of the IUPAC Commission on Macromolecular Nomenclature and a Consultant to the IUPAC Sub-committee on Solubility Data. In addition Dr Loening is a member of the International Federation for Documentation Committee on Universal Decimal Classification in Chemistry and Chemical Technology and is Consultant on drug and pesticide nomenclature for a number of other organisations.

Professor N. Lozac'h studied at the École Normale Supérieure, Paris and held posts as lecturer at the Universities of Lille and Paris. In 1952 he was offered a Professorship at the University of Caen. He became Director of the École Nationale Supérieure de Chimie, Caen, from 1952 to 1977, Dean of the Faculty of Science from 1957 to 1969, and since 1977 has been the Director of the Institute of Material and Radiation Sciences at Caen. Professor Lozac'h is an associate member of the IUPAC Commission on the Nomenclature of Organic Chemistry and was its Chairman for six years. He is also a member of the Commission française de nomenclature en Chimie Organique and has variously held the post of Secretary and Chairman of this Committee.

Professor M. F. Lynch studied organic chemistry in Ireland, West Germany and Switzerland. Subsequently he joined the Research Division of the Chemical Abstracts Service and spent four years researching chemical information systems. He has been at Sheffield University since 1965 where his primary interests are in computer systems for information handling.

Dr D. S. Magrill studied at the University of Glasgow and was a postdoctoral fellow at the Prairie Regional Laboratory, Saskatoon, Canada, working on biosynthetic studies of flavonoid compounds. Later, as research fellow at the Technion-Israel Institute of Technology, Haifa, he worked on propellanes and in the field of X-ray crystallography. Dr Magrill joined the Beecham Group to work on physico-chemical studies of the ingredients of toothpaste. He later became involved in information work concentrating on the design, evaluation and implementation of computer-based information systems.

Dr G. P. Moss is a lecturer in organic chemistry at Queen Mary College, London. His particular research interests are in biosynthesis, synthesis and spectroscopic properties of the carotenoids, vitamin A and related natural products and their analogues. He is Chairman of the IUPAC Commission on the Nomenclature of Organic Chemistry, Convenor of the Working Party on Stereochemistry, a member of the IUB-IUPAC Joint Commission on Biochemical Nomenclature and a member of IUPAC working parties on natural products and fused ring nomenclature. Dr Moss is also a member of the Joint Nomenclature Panel of the Royal Society and Royal Society of Chemistry.

Mr M. G. Robiette has worked for the Royal Society of Chemistry (RSC) since 1967. His early career was concerned with the production of primary journals for this organisation. Following a year's secondment to CAS in the United States, he transferred to abstracting and indexing material published in the United Kingdom for CAS. Mr Robiette is currently Manager of the Abstracts and Index CS Department of the RSC and has responsibility for the production and development of secondary data bases in areas of chemistry, mass spectrometry and analytical science.

Mr D. Rossitter is currently Trade Marks Counsel for Lilly Industries Ltd, a subsidiary of the American pharmaceutical company, Eli Lilly and Company. In this capacity he is responsible for trade mark administration in Europe, the Middle East and a large part of Africa. Mr Rossitter was called to the bar in 1968 and has been Chairman of the Pharmaceutical Trade Marks Group for over ten years. He is a member of a number of intellectual property committees including those of the Bar Association, the Association of the British Pharmaceutical Industry, the combined working party of the Law Society and the Bar, the Trade Marks Patents and Designs Federation and the European Federation of Pharmaceutical Industries and the UNICE. Mr Rossitter has been co-opted as adviser on the agricultural pesticides nomenclature committees of the BSI and International Organization for Standardization (ISO) secretariates.

Mr T. C. Swinfen has been Senior Science Master at Uppingham School since 1966. His earlier teaching career was in the Royal Navy, at Rugby School and at the Lake College School, Ontario, Canada. Mr Swinfen is a member of the Joint Nomenclature Panel of the Royal Society/Royal Society of Chemistry and is a former member of the Tertiary Publications Committee and the British Committee on Chemical Education. He is an active member of the Association for Science Education and serves on the Education Co-ordinating Committee, the Publications Committee, the Working Party on Chemical Nomenclature and is Chairman of the Chemical Labelling Sub-Committee. Mr Swinfen is Assistant Editor of *School Science Review.*

Dr W. G. Town undertook research in crystallography prior to joining a team at the University of Sheffield working on information systems. After a period spent at the Cambridge Crystallographic Centre, Dr Town moved to the European Community Research Centre at Ispra in Italy to work on the development of a data bank for environmental chemicals.

Mr R. B. Trigg worked in the research laboratories of Glaxo and Twyford Laboratories prior to joining the secretariat staff of the British Pharmacopoeia Commission (BPC) in 1971. At the BPC he has responsibility for work on approved names and general analytical methods. With colleagues, Mr Trigg maintains the upkeep of the many BP monographs on synthetic, natural and inorganic materials and their preparations. He serves on the WHO panel of consultants on pharmaceutical names and for a number of years has been a member of the United Kingdom delegation to the international plenary meetings of the International Standards Organization Technical Committee on pesticide nomenclature.

Mr S. B. Walker is an information officer specialising in chemical information at the ICI Plant Protection Division at the Jealott's Hill agrochemical research centre in Berkshire. On joining the company he worked on research on synthetic chemicals, relating to pharmaceutical products but then specialised in chemical structure searching and indexing using computerised techniques. His current work is concerned with the acquisition, indexing and searching of chemical substances for a large data bank relating to the work of his organisation. Mr Walker has been an active member of the United Kingdom Chapter of the Chemical Notation Association (CNA) since its formation in 1969 and is a member of its executive committee and tutorial organiser. Working with representatives of other leading United Kingdom pharmaceutical companies, Mr Walker has developed a data base of commercially available chemicals known as the 'Fine Chemicals Directory'.

Dr S. E. Ward is Manager of Science Information at Glaxo Group Research with responsibility for library and information services to research and development staff at four sites. She specialises in the computerised handling of chemical structures and in techniques for the acquisition, storage, retrieval and communication of physico-chemical and biological data. Dr Ward was Chairman of CNA at the time of the Symposium and a member of the Education Committee of the Institute of Information Officers. She is now chairman of the Publications Committee of the Institute of Information Scientists.

Dr W. A. Warr studied chemistry at the University of Oxford and undertook doctoral research on the reaction of azides with indoles. Whilst undertaking part-time work at the Oxford University Experimental Information Unit she became one of the first English users of the Wiswesser Line Notation system. Dr Warr is now a Senior Information Scientist and Systems Analyst at the ICI Pharmaceutical Division at Alderley Park, Cheshire. At the time of the Symposium she was President of CNA's United States parent organisation and a member of the Joint Nomenclature Panel of the Royal Society/Royal Society of Chemistry, and the BSI committee on chemical nomenclature.

Mr A. G. Wyatt qualified at Manchester University College of Technology. He gained extensive industrial experience through work at ICI Ltd, Pyrene Company Ltd, and Toni Cosmetics Ltd. Mr Wyatt joined the Association of British Chemical Manufacturers in 1959 and stayed with the organisation when it became the Chemical Industries Association. He was appointed Secretary of the British Colour Makers Association in 1973, and has served on many committees representing the chemical industry. He is currently chairman of the relevant committee of BSI working on approved nomenclature for chemicals of interest to industry.

The opinions expressed by the authors of the papers included in this book are their own and do not necessarily represent those of their organisations.

CHAPTER 1

Problems of use of chemical nomenclature in specialised areas

---

(i) CHEMICAL INDUSTRY

A. G. WYATT
Chemical Industries Association, London, United Kingdom

This short contribution indicates some of the problems in chemical nomenclature which have arisen in industry and some of the attempts made to cope with them. The main chemical trade association in the United Kingdom now called the Chemical Industries Association (CIA), has always been concerned with the use of appropriate chemical nomenclature. For many years the Association has published a directory called *British Chemicals and their Manufacturers.* The current version of this directory is sold by CIA under the title *Chemicals* with the year of publication e.g. *Chemicals 80*.

The chemical names in this book have always followed the recommendations of the British Standard (BS) 2474, the industry standard on chemical nomenclature prepared by the British Standards Institution (BSI). The committee responsible for BS 2474, *Recommended Names for Chemicals used in Industry* is designated CIC/3. This committee includes amongst its membership representatives from the Laboratory of the Government Chemist, the Royal Society of Chemistry, the Pharmaceutical Society and industrial companies. It meets as and when required to consider whether the current standard needs revision and if so in what way.

The Standard, first produced in 1954, was subsequently revised in 1965 and currently is undergoing further revision. The draft of the new standard is expected to be circulated in 1982.

The main problems which have arisen in the chemical industry are:

(a) the reluctance of personnel in chemical companies to give up using old, archaic or misleading names and even acronyms for the chemicals they are using every day in their work. For example, names like caustic soda, oil of vitriol and, even not so far back, muriatic acid. The last was shown in at least one company's work sheet as MA2. There are also acronyms like MEK, BHC, etc.

(b) the development of an appropriate system (not too complicated) that can be sold to industry for their use and which they will use. This was one of the main aims behind the creation of a British Standard giving recommended names.

The Standard has gradually gained acceptance amongst users and in the chemical industry and is now widely followed in the United Kingdom. It has always been sponsored by the Chemical Trade Association (now CIA), the Laboratory of the Government Chemist and other bodies. Examples where the philosophy adopted in this Standard has been used are:

- the Association's directory;
- Tremcards, manuals on labelling, etc. (published by CIA);
- the listing of chemicals in legislative instruments produced by Government, e.g. the list of temporary exemptions (pre-EC), transport regulations, etc.

There are examples where specially contrived and/or coined names are used and officially approved in describing chemical substances. These systems were established in an attempt to solve problems where long and/or complex chemical names existed for materials such as pharmaceuticals and pesticides, and where, in the industries concerned, it was the end use of the products rather than their chemical structures that mattered. Examples are the British Pharmacopoeia Codex names for drugs and medicines, pesticide approved names such as WHO, INN. Undoubtedly these systems are very useful in the context in which they are used. It is for example much easier to talk of aldrin, monuron, chlorothiazide and butazolidine, rather than to use the full International Union of Pure and Applied Chemistry (IUPAC) names for such materials. This form of usage of course is not really nomenclature in the true sense, but nevertheless it has become widely accepted in practice, for example in the medical profession and the agricultural industry. At one time these names were used in legislative instruments such as those listing temporary exemptions.

Again experience has shown that in particular the laboratory supply houses have been very good supporters of the Standard in the labelling and cataloguing of their products for sale, on the other hand heavy industry has perhaps been less diligent in using its recommendations. However, in recent years, particularly post-EC, with all the various legislation, schemes for labelling of dangerous goods for transport and other purposes, etc., uniformity and adherence to a system has become more widespread. In this area the United Kingdom has always tried hard to persuade others to follow BS 2474, and therefore largely the IUPAC system, but it has not always met with success. In fact, it has been noticed amongst contacts in Europe that all the European bodies do not necessarily harmonise themselves in the sense of the nomenclature they use. Some transport regulations for example use different names compared with, say, dangerous substances legislation. The main need in systems for industrial use is that they should be relatively uncomplicated and easily reproduced and printed for use by companies.

Indexing for reference purposes can be a problem, particularly in the case of directories, buyer's guides and similar publications. For many years the BSI committee supported italicising all prefixes such as iso-, sec-, tert-, sym-, and trans- as an aid to indexing. In the face of continual pressure from IUPAC and from the United States of America, BSI has changed its position on this matter and will be recommending in its new draft standard the romanising of 'iso-'. The problem of indexing chemical names can lead to some unusual situations. A telephone enquiry directed at the CIA requested information on the material 'green oil'. The inquirer said that he had been unable to locate references in the Association's directory. The material was listed, but under 'oil, green'.

One of the main differences which exists in the usage of chemical nomenclature is the distinction between the sort of problems that occur in industry and those which occur with many of the other bodies. By and large industry is not too bothered about the actual name given to specific chemicals, be it phenol, carbolic acid or hydroxybenzene. However, most of the other bodies concerned with this subject are preoccupied with the problem of deciding on a particular description for a particular chemical when four or five good and acceptable alternatives may exist. One criticism of IUPAC, is that it does not generally opt for a preferred name when several options are given. The work undertaken by BSI does list a preferred name in every case and industry welcomes this kind of guidance.

Not only industry but all concerned with chemical nomenclature would gain considerably if an international agreement could be reached giving unanimity on the name which is to be given to any particular chemical material regardless of language and any other considerations. This would be of enormous benefit in trade, commerce, learning, research and any area in which chemicals are used and chemical names appear.

(ii) EDUCATION

T. C. SWINFEN
Uppingham School, Leicestershire, United Kingdom

Chemistry, for the ordinary mortal, is a very difficult subject. To become good at it involves not only the learning of a new language but also a great conceptual jump. The student starts with the joyful qualitative cookery of the young – growing crystals, boiling liquids, producing colour changes and smells. To go from here to the *science* of chemistry and a mastery over chemical changes involves comprehending the concept of the molecule and realising that chemical reactions are a manipulation of invisible atoms and molecules or of abstract symbols and formulae. This is a very hard step to take and it is these problems faced by the student during and just after his or her school career which are considered in this paper. Recently sixth-form students were asked when they thought they had taken this step. Most said 'during the O-level year', some up to a year earlier. [*Editor's Note.* O-level 15/16 years old; A-level 17/18 years old.] These are bright students – some will win Oxbridge scholarships.

Imagine the difficulties of a teacher who is trying to relate events in a test tube to changes at a molecular level for students not clever enough for university – or for O-level or A-level. His needs should not be judged against the experience of a skilled nomenclature expert. The use of appropriate nomenclature may be an important factor in the ultimate success achieved by the student. People move from job to job more often than they used to and their families of course go too. A system of nomenclature is needed which does not impose a new vocabulary every time the student changes text books, Examination Board or school. Since 1972 all text books used in schools and all Examination Boards in the United Kingdom have used the same chemical names.

The 1960s saw the development of the Nuffield courses. In 1967 the *School Science Review* – the journal of the Association for Science Education (ASE) – carried an article giving the list of names used in the Nuffield O-level chemistry course. This provoked correspondence and led to the setting-up of the ASE Working Party on Chemical Nomenclature. This Working Party had three possible courses of action, (a) to do nothing much, (b) to make minor changes and (c) to think out the schools' problem from first principles. In the event it chose the last course. Through correspondence, journals and public meetings, its members consulted science teachers, IUPAC Commissions, GCE Examination Boards and anyone else who seemed interested and could be reached.

The Working Party's decision was that each substance should be assigned one name and one name only. In a very few cases dual-naming was recommended (e.g. acetic acid). IUPAC permits alternatives but in schools their use would cause confusion. A French child learning English may learn the word 'cat'. The words 'pussy', 'pussycat', 'kitty', 'moggy', and so on come later, if at all. One animal – one name. Secondly it decided to publish a list of names to be used in schools, rather than a set of guidelines for the individual teacher to interpret.

What is the purpose of a chemical name? It can be used to identify what is in a bottle. A trivial name such as methyl orange does this. The name can be used to give the molecular structure of the compound – for example, propan-2-ol. The Working Party decided that if the student needed to know the structure of a compound then its name should be fully systematic, constructed according to IUPAC principles on a substitutional basis. CH_3Br is named as bromomethane, not 'methyl bromide' and certainly not both. If the structure is not needed then the name could be a trivial one.

A further problem relates to elementary work. It would obviously be unrealistic and over pedantic to use the name copper (II) sulphate-5-water in a class of eleven-year-olds. Children will probably refer to the substance as 'that blue stuff' and for them that is a perfectly good unambiguous name. When they know that it contains copper, the name copper sulphate begins to make sense. An understanding of ions and ionic charge causes the name to become copper (II) sulphate even though oxidation numbers may be a couple of years in the future. Water of crystallisation can be included in the name when it becomes relevant. A commonsense progression of this type was described in a Report of the Working Party nine years ago. Nevertheless the inevitable zealots ignored it and went prematurely and rigidly systematic, causing difficulty for the pupils and generating adverse criticism.

It is a surprising fact that a particular compound may be met quite infrequently during a school chemistry course. For example, phthalic anhydride appears in an A-level course only twice: once in the fluorescein reaction, by itself, and once associated with the sole appearance of phthalic acid.

There is little point in learning the trivial name when it is the formula the student needs to know. From the formula the systematic name follows easily, or vice versa. The breadth of the problem can be envisaged by considering the need to learn the formulae of compounds at a single meeting of each of some 100 trivial organic names including, for example, cumene, oleic acid and *p*-toluidine.

A quotation from a letter written by a senior member of a large company on this subject to the Working Party sets out this problem. 'I would merely say, from my own experience, that the illogical morass of trivial names that have been used in organic chemistry imposes an unnecessary burden on the memory and for some pupils must be a serious distraction to learning. I believe that it would be far easier to associate trivial names with a logical nomenclature if that were learned first and so I do not see the recognition of trivial names, when they are still used in industry, as a real problem – each special case could easily be related to the formal name already learned.'

So far the considerations have been concerned with the needs of the student, but students grow up and may become chemists or workers in a chemical plant. There are three interfacial problems and these can be over-simplified by labelling them 'industry', 'higher education' and 'research'. In real life these interfaces overlap. Oversimplification is a pedagogical technique and this contribution has been written by a teacher.

The difference between the needs of industrial nomenclature and those of education is that in industry the name is employed to label the container and in education it serves to teach the molecular structure of the contents. For industrial purposes a simple unambiguous trivial name will do and the last thing anyone wants is innovations which will upset the work force – 'That propanone stuff's rubbish – give me acetone every time'. The Working Party discussed this in the first edition but in the second edition laid greater stress on the need to teach both names once the chemical principles taught by the systematic name have been understood. Teachers have not done this enough in the past. Even so the teaching profession cannot anticipate every need of industry and should not have to use such names as 'crushed blue vitriol, 98% pure' or explain the properties of DOV. During a recent visit to a fertiliser factory, staff were asked which alkali was produced by the potash plant, potassium carbonate or potassium hydroxide. 'Neither', they chorussed, 'oxymuriate of potash'.

By the higher education stage, the student will with luck have left school qualified to study molecular sciences by actually understanding the principles of chemistry. (It should be remembered that an E grade in A-level chemistry probably represents a mark of less than 50%.) Teachers must prepare their students for higher education, as for industry, by introducing the commonly accepted names towards the end of the course. On the other hand for compounds which appear only rarely in a A-level course this might not always be practicable, and therefore lecturers in higher education need to be aware of what use schools are making of systematic nomenclature, with some understanding of the reasons. They will then be able to treat the student sympathetically and give some help in bridging the gap. If the student has been taught properly he will quickly adapt. Solving this problem is like building a large bridge; start at opposite sides of the gap but meet in the middle.

The chemical literature is written using the traditional variety of names and the student has just got to use them too. If he or she has been in higher education only a short time some help will be needed. The Working Party kept in touch with developments in Chemical Abstracts, etc., through the Royal Society of Chemistry.

It must not be forgotten that the research field has its own jargon with sets of meaningful initials, etc. This could not be taught in school courses whatever system of nomenclature was used.

Summary

To the beginner in chemistry a limited and simplified vocabulary, much more rigidly systematic than it used to be, is an aid to an understanding of the subject. It may not necessarily be appropriate to the later needs of the few whose careers lie within chemistry.

There could be a confrontation between, on the one side the 'zealots of the utterly systematic nomenclature' and on the other the 'let's all go back to 1955' reactionaries. Such a confrontation is implied by the use of terminology such as 'realistic' and 'systematic' nomenclature. For much of the chemistry taught in schools the two are the same.

In a British Committee on Chemical Education survey made while the second edition of the ASE Report was approaching publication, comments on the draft Report were mixed, but on the whole adverse. However, 74% of the respondents had not seen the Report before, including 61% who had not even known that it existed. It is not confrontation that is needed now, but co-operation. After all, any scientist knows that according to the Gaussian distribution curve there is, between the two extremes, a great deal of middle.

Note. Copies of the Report referred to in this paper may be obtained from The Association for Science Education, College Lane, Hatfield AL10 9AA, England.

(iii) SCIENTIFIC PUBLISHING

M. KENWARD
New Scientist, London, United Kingdom

Most journalists on the editorial staff of technical and scientific magazines, those who have to deal each day with complex articles, try to avoid the use of complex chemical nomenclature wherever possible. This is not from a belief that it is unimportant to put the right names to chemicals but because most publications, even within the field of scientific publishing, are not written solely for the research chemist. Given an unlimited remit most research chemists would litter the pages with complex nomenclature that would discourage immediately many of the readers from other disciplines. The limitations placed on the use of chemical nomenclature is not a wanton disregard for the feelings of chemists but an acknowledgement of the facts of life of publishing. Printers also find it extremely difficult to reproduce accurate nomenclature given the speed of publication and this is an obstacle that must be kept in mind by editorial staff before using a complex system for describing scientific matters.

Chemists spend much reading time in perusing the pages of the specialist press, but many of the magazines that are read devote most of their efforts to reaching a wider audience. The judgement of the use of nomenclature by a publication must take account of its audience. Chemists should take note of that old adage from Gilbert and Sullivan 'to make the punishment fit the crime'. To put this expression into a framework that makes it relevant to the naming of chemicals the words should fit the context or, alternatively, the nomenclature should fit the audience. The weekly journal *New Scientist* probably reaches the most general scientific audience that can be envisaged and yet it is not typical of a scientific publication. On average 85,000 copies are sold each week and local circulation lists within establishments probably raise that figure to a reading public of more than half a million people. Chemists tend not to consider the general scientific community when using nomenclature, but articles on chemistry need to be readily understood by physicists, engineers, biologists and perhaps even the odd politician and economist who reads *New Scientist*. These comments apply equally to other general scientific journals such as *Science* and *Nature*.

A physicist is not likely to be widely interested in an obtuse and completely accurate description of a chemical compound if that physicist's chemical education did not equip him with the fundamentals of nomenclature. Journalists and editorial staff involved in general scientific publishing tend to retain only the minimum of the detailed chemistry from their student days. The same is true of many readers of scientific journals, their chemical education ended long before chemistry teachers attempted to inform pupils of the esoteric art of naming chemicals.

The use of nomenclature in a journal must therefore be judged against its audience. A chemist who wises to contribute to a journal must consider the type of readership before sitting down to write an article. The editorial staff of journals such as *New Scientist* can be aligned with interpreters of foreign languages who have to translate chemistry into a language that a physicist can readily understand. By the same definition staff also have to translate physics into a language that the chemist can understand. In addition the change had to be done in a way that keeps the reader interested in the subject.

Obviously any translation is helped immensely if the original writer does much of the work prior to the submission of the article. If an article is being prepared for a wider audience, be it the readers of a general scientific journal, or perhaps a daily newspaper, or maybe a scientific paper that will reach a very wide audience, then it should not be left to the editors to modify the text. An author should query the need for a long and complicated chemical name and ask whether it could be simplified? If the answer to any of these questions is 'no' then is it possible to add extra information that would help or would assist the non-specialist to understand the subject? A similar approach should be adopted when writing for the specialist literature. However, the journals in question may dismiss something that appears too easily understood on the grounds that if it is that simple, then it cannot be very good. Whether such an attitude applies is difficult to assess, although some journals give the impression that the harder a paper is to understand then the easier it is to achieve publication. Any journal that oversimplifies to the extent of getting the subject matter wrong cannot be defended. There are a number of examples which illustrate how publications can get facts wrong when writing about chemicals. This does not relate to the misuse of nomenclature but the misuse of plain English and perhaps the misuse of the truth.

There is the common error, when writing about the use of tetraethyl lead added to petrol, of referring to the use of lead. This is a misleading usage but in most contexts it could be argued that it does little harm to allow the public to think that there is a high level of lead in their petrol. A more damaging misuse of the language appeared in a well-known political weekly. The article was concerned with an incident where dozens of children at an outdoor concert suddenly became ill. Experts claimed it was mass hysteria whilst others attributed the incident to impurities carried in the air. The political weekly undertook investigative journalism and headlined its findings on the front cover as nerve gas. It might be assumed that the army carried out secret tests with some new agent of chemical warfare. However, the nerve gas suddenly underwent a chemical transformation in the article carried by the magazine and it became a pesticide. The relationship of the pesticide to a nerve gas was claimed on some slight similarities to a chemical product that is of interest to the military. It was by this obscure route that nerve gas ended up on the front cover of the journal. The problem of getting the right name for the chemical seems less important when mistakes of this magnitude take place. It is not clear whether the weekly was being deliberately provocative or whether the journalists did not understand the subject matter. It is this type of chemical error which is far more annoying in journalism than an honest mistake when naming a chemical completely accurately. However, mistakes of this type can produce amusing headlines. A story that appeared in an American magazine on United States plans to make gasoline from coal was headlined, 'The Explosion of Coal Liquifaction Projects'.

Other branches of science use jargon in the same way as chemical nomenclature and authors expect magazines and journals to get the usage in the correct form. There is what could be termed physical nomenclature. Physicists continue to think long and hard about descriptions of the physical world. Whenever a physical unit is printed there is always a reader who will write to say that the words chosen were in last year's style.

Perhaps this underlines one of the basic problems that must be faced in scientific publishing. Scientists themselves keep changing the rules on the way things are to be done. As soon as editorial staff have come to grips with one set of units then some international or national organisation decides that a new system is required.

The physical sciences such as chemistry and physics are not the only specialist fields with a language that publishers are expected to observe. Biologists use strange Latin names in their papers and a journal has only to make one mistake when using these to bring a flood of letters. It is not being argued that scientists should make major allowances because a publishing house has to deal with these problems. However, when a mistake is suspected then the first question should be whether it is an error or an attempt to achieve simplification in order to make life easier for the reader. Any contribution that includes references to chemical nomenclature should be read sympathetically and written thoughtfully. The use and misuse of nomenclature is in the hands of the chemist.

Papers that are littered with detailed chemical descriptions may not be appropriate for general audiences who read particular journals. Papers intended for other chemists can be as obscure and incomprehensible as is desired. A paper on a scientific topic that is so important that it has to be understood by physicists, environmentalists and perhaps even journalists should be carefully prepared in relation to the words that are used. There is nothing wrong in using nomenclature when writing on an important subject that has to be brought to the notice of a wider public. Here the same principle applies as when dealing with contacts from the press; the information should, for sake of the subject, not to mention the accuracy of the press story, be clear and simple and the reporter should not be baffled by the use of scientific jargon. If the simplification is chosen by the scientist then the journalist cannot go away and turn a pesticide into a nerve gas. Equally there is nothing wrong to the use of this approach when writing for equals. It is an error to baffle others with confusing nomenclature. An author should identify what his colleagues need, draw back and sympathise through his use of words. There is nothing wrong in making the same point in a number of different ways if this helps to simplify a subject. The important issue is not trying to prove how clever the author is but how easy it is to understand his complex specialist subject.

(iv) PATENTS

C. D. KOPKIN
Patent Office, London, United Kingdom

Nomenclature problems experienced with patents are probably very different from those in other areas although there is undoubtedly overlap with information science.

The importance of nomenclature in patents is two-fold. Firstly the scope of a granted patent has to be clear and precise, and secondly the invention has to be new and non-obvious. This implies that when an Examiner carries out a search he must be able to find, both in earlier patents and in literature generally, the same chemical as, or chemicals that are very similar to, the chemicals that are the subject of or used in the invention. One important factor is that the Patent Office has no control over the terminology or nomenclature that is used by the applicant in the patent application when first filed. This is particularly important under the new Patents Act in the United Kingdom, and of course under the European Patent Convention, because in both cases the initial application is published, in so far as it is at all possible, in precisely the form in which it is filed. This includes any and all imperfections – imperfections of style, spelling mistakes, chemical mistakes and in particular, as far as this book is concerned, imperfections in nomenclature. The point to note is that this document, which is published and, as mentioned, may contain imperfections, is a search document for all future patent searches. This particular point will be returned to later.

It is found that problems in nomenclature arise for a number of reasons. One example is that of the same chemical being known at different times and/ or in different documents by different names, for example acrylonitrile-vinyl cyanide. In organic chemistry this problem is met to a considerable extent in the Patent Office by classifying and indexing the structure of the chemical and not its name or the name given in a particular patent application. In other chemical areas, there may be no, or perhaps just a rudimentary, classification of the chemicals and the problem of recognition is there much greater. In any case Examiners must be able to recognise a particular chemical under all its possible names, because in the great majority of cases it is the name that appears in a Patent Application, Patent Document, and in the literature generally.

A second problem in interpretation of patents is the use in patent documents of terms in a sense that is different from the usual or even the dictionary sense. For example 'vinyl polymer' has been used to include any ethylenic polymer; 'alkyl' to include cycloalkyl, chloroalkyl or substituted alkyl generally.

What is important here is that subject-matter cannot be added to a patent application after it is filed, and the problems this causes may be illustrated by a simple example. If the claim in a patent under examination has 'alkyl', yet included in the examples are some which have say vinyl or allyl in the position where that alkyl should be, what can the applicant be allowed to put in his claim? He is entitled to protection for those particular compounds even though they are not strictly within the claim, and he has to be allowed to amend his claim to bring them into its scope; so how can he do this? Can he put 'alkyl which may be unsaturated'? No, because that brings in acetylenic compounds and there is no basis for these at all. Even 'alkenyl' may be too broad, depending on the circumstances, because this brings in a complete new class, a new range of chemicals, on the basis perhaps of just one or two very specific examples. There are all sorts of possible variations of this particular type of problem and each has to be dealt with individually on its particular merits. There is no general answer.

A third type of problem is the use of ambiguous or imprecise terminology or nomenclature. If the word 'lime' turns up in a document found in a search, what does it mean? It might be slaked lime or it might be quicklime. What is the scope of a claim which requires the use of 'a transition metal' as a catalyst? What precisely is a 'noble metal'? Does 'alkaline earth metal' include magnesium? Different users of these various terms often mean different things by them, and even different dictionaries can and do give different definitions of them. Returning to organic chemistry – take phenyl acetamide – this could, of course, be the amide of phenylacetic acid, but it could also be the anilide of acetic acid, and unless there is some distinguishing feature given, which one it is is unknown. Thioacetic acid methyl ester may be the O-methyl ester or it may be the S-methyl ester. Where is the chlorine in chlorotoluene?

It may be said that these are not problems of nomenclature as much as problems of bad nomenclature, and this, of course, is correct. But unfortunately these problems do occur, both in patent applications being examined and in documents being searched. In these situations comparison may be very difficult and therefore inventors and patent agents should be as precise as possible in the chemical nomenclature used – at least when filing patent applications. Not only would this make the Examiner's job easier, but it really would be in the claimants' interests also, because, in fact, if there is something in the original document which is at all vague or ambiguous it may not be possible to allow correction later if this correction would result in specifying something which was not specified before the correction or amendment was made. This is part of the bar against adding subject-matter to a patent application after its date of filing.

Returning to methyl thioacetate as an example, this name specifies neither the O-methyl ester nor the S-methyl ester. Unless there is evidence in the original document, for example, a general formula, as to which form was really intended, the applicant may not be allowed to write that form in later. A further consequence of this is that because of the very stringent requirements for demonstrating lack of novelty in a patent, it may well be the case that if somebody comes along later specifying the use of say the S-methyl ester in an invention of a very similar nature, then because S-methyl has not been specified earlier the new application may be allowed.

(v) INFORMATION SCIENCE

Professor M. F. LYNCH
University of Sheffield, Sheffield, United Kingdom

In this brief contribution both a backwards and forwards look at the influences under which nomenclature has developed will be considered, particularly in terms of the functions which it serves. Consideration will also be given to the effects which new technology, now rapidly becoming available for working chemists, is likely to have on the requirements for nomenclature systems. The comments deal mainly with the generality of organic chemical structures.

Nomenclature is an integral part of communication in chemistry and allied disciplines, a tool both for those who are expert in structural chemistry, as well as for those who are inexpert but are dependent on names for chemical substances. Nomenclature is, accordingly, a family of languages, its members serving, in various ways, the needs of both expert and inexpert users and also the needs of specialist subgroups. Hence, nomenclatures are highly dependent on the contexts in which they are employed. Expert and non-expert users have very different requirements in terms of their relative emphasis on structural characteristics, i.e. the need for a largely explicit structural representation. Problems therefore arise particularly at the interface between expert and non-expert usages.

The need for a fully explicit and systematic representation of a chemical substance, so that its presence or absence in an index can be readily determined, has long been a dominant factor in the design of nomenclature systems. Computer and telecommunications technology, evidenced particularly in the Registry System and in CAS Online and its potential extensions, will provide alternative means of fulfilling this function with greater reliability, as well as permitting the retrieval of any or all recorded names for existing substances, together with indications as to particular preferences or standards.

The second dominant factor in nomenclature design is the need to highlight 'interesting' structural features, so that forms of relatedness among groups of substances may be made evident however these forms of relatedness may be perceived by the chemists concerned with these classes of structure. This is closely allied with the need for compactness and memorisability in chemical names. Focussing often on ring systems, but by no means exclusively so, as the prostaglandin names show, much of the structure is described implicitly, rather than explicitly. This reflects a natural tendency in the use of any language – a striving for economy in naming entities which we encounter frequently.

The need to reveal higher-level forms of relatedness causes substantial problems when low-level searches, exemplified by substructure searches, to identify classes of molecules containing similar constructs defined at the level of atoms and bonds, are needed. The implicitness of the higher-level description tends to mask structural characteristics at the micro-level; this is also the case with chemical line notations. Thus, for example, substructures which may occur wholly in cyclic portions, wholly in acyclic portions, or arbitrarily between the two, cause especial problems when nomenclature is used as the search medium. Another example is given by forward or backward citation even of simple substructures, e.g. phenoxyethyl versus ethoxyphenyl. Although nomenclature search systems can provide substantial results in support of substructure search, to meet this requirement fully calls for different methods, notably the CAS Online and the DARC substructure search systems. Is it wise to see nomenclature in the future as the appropriate medium to support this requirement?

So far, this view of chemical nomenclature in relation to information science has seemed to predict a steady erosion of the functions fulfilled by it. It is certainly true that graphics terminals, with on-line links to data bases, will feature much more strongly in the laboratories of those for whom structure is important, while access to data bases of synonyms is already widely available for those for whom properties are more important. Again, one must keep the availability of automatic translation from nomenclature into other computer-orientated descriptions in mind.

What characteristics does nomenclature have that are not readily supportable, as yet, by the available technology? Nomenclature has an unrivalled capacity for generalisation of groups of molecules with certain similarities. We readily understand expressions such as octadienones, diarylpyrazolines, azepinoindoles, for classes of molecules which are not readily represented by structure diagrams. Structural chemistry is largely supported on the foundation of such generalised descriptions, and their importance in patents is considered by C. D. Kopkin in Section (iv).

It may appear that a kind of anarchy has been advocated but further developments in nomenclature must reflect strongly the context-dependency of the use of sub-languages of nomenclature, and their reflection of the many different specialisations. Thus the preservation, indeed the enhancement, of substantial degrees of freedom and flexibility in nomenclature is advocated. This is especially so for research level use so that, freed by the new technology from certain of the constraints which have influenced it in the past, it may flower yet more strongly in order to enhance both written and spoken communication in support of structural chemistry. Horses for courses, the appropriate tool for the job in hand, chemists have always used them and will continue to do so.

This goal of much more responsive information systems will be reached only in the long term, but the beginnings have already been made. Clearly, there is the problem of managing change in the intervening period, and of ensuring multiple modes of use for a substantial time in the interim, so that the goal of devising better systematic nomenclatures needs to continue. In this respect much new knowledge on the theory underlying the design of formal languages has been gained, particularly from the design and compilation of modern programming languages which can be applied usefully to this work.

CHAPTER 2

Problems in the day-to-day use of chemical nomenclature

E. HYDE
Fraser Williams (Scientific Systems) Ltd, Poynton, United Kingdom

---

Chemical nomenclature has always been the overworked and underfed carthorse of the chemical profession. Too much has been expected from it and too little attention has been paid to it. Nomenclature is expected to perform a number of tasks.

(a) Registration – for which a unique name is required.
(b) Generic classification.
(c) Description of reaction routes.
(d) Verbal communication.
(e) Capability of machine interpretation.

When nomenclature has shown a tendency not to fulfil an immediate need then the answer has been to find another way to fulfil that need. Yet no other method of describing molecules can take over all the functions of nomenclature.

The problems of nomenclature are more fundamental than its use in different languages. They are caused by the diverse roles expected from chemical names and the failure to structure nomenclature in a manner more suitable to the tasks that have been assigned to it.

Chemical nomenclature has evolved rather than been systematically applied to the solution of any problem. The most gross mistake in the use of nomenclature has been to assume that one name can be devised which would satisfy the day-to-day needs of the chemist and the non-chemist and the ambitions of the information profession.

The current problems facing chemical nomenclature can be considered under three headings: first those relating to the people who create chemical names, the chemists and the indexers, then the problems facing the users of chemical indexes and finally ways in which present-day indexing practices could be improved by the logical structuring of a chemical name and by reducing the volume of compounds to be indexed.

CREATING CHEMICAL NAMES

The chemist

A chemist needs to describe a compound, in a chemical paper or at a chemical meeting, in terms of the specific chemical problem he is currently studying. If he has been preparing a series of compounds then he will devise names for all members of the series such that the relationships and essential differences between the compounds are reflected in the name. Providing names for compounds in a patent specification often requires a similar approach, for example when it is essential to retain a radical name associated with the property claimed. The name chosen by the chemist or the patent officer reflects the current role of the compound.

The indexer

Names devised by the chemist cannot always satisfy the needs of an indexer, who requires a unique name for each compound. Rules for developing a systematic name have emerged, but these have often been devised in response to a problematic compound rather than in an attempt to devise a system. The indexer should consider the needs of retrieval of information about a compound. It is not sufficient for the indexer just to create a unique name to appear in the correct place within the index. He can also set up rules to ensure that classification by specific groups is reflected in the name.

Over the years indexers have considered a number of classification methods:

(a) by structural type, as in Beilstein;
(b) by molecular formula;
(c) by reactions;
(d) by subject;
(e) by inverted names.

Inverted names were devised in an attempt to use nomenclature in the secondary role of generic classification so that analogous compounds would be filed adjacent to one another in an index. Recently indexers have considered that machine interpretation of names is an economic necessity and hence have removed many names outside the vocabulary of the practising chemist. One of the reasons for the difference between a name produced by a chemist and one produced by an indexer is that the chemist's name must be suitable for both written and verbal communication. The indexer's name is only required for print purposes.

THE USERS OF CHEMICAL INDEXES

The chemist

The day-to-day needs of the chemist can be considered under three groupings.

(1) Verbal communication.
(2) Visual communication.
(3) Searching for: specific compounds;
reactions;
generic groups;
sub-structures.

A chemist assigns a name to a compound and is then able to use that name for verbal and visual communication. However, when it comes to searching for compounds using nomenclature, he is in the hands of the indexer who has created the names. An index composed of systematic names cannot satisfy all the search requirements of the chemist no matter how ingenious the deviser or indexer: the classification schemes imposed on top of rule changes of the various systematic nomenclature systems have alienated the chemist. He moved to molecular formula indexes many years ago. Systematic nomenclature is a lost art to the practising chemist. To him, systematic names are located by searching for them in a secondary index, such as a Molecular Formula index. Systematic nomenclature is now entirely in the province of the information profession. It is not essential to the chemist.

The non-chemist

It is various services based on chemistry which rely entirely on chemical names: the chemical manufacturers, other scientific professions, medical workers, environmental controllers, the legal profession, chemical transporters and store-keepers. These people use nomenclature to enable them to access essential data and they require the support of the chemical profession along with the information workers. However, the chemical profession is not addressing itself to this problem. The chemist has played little part in the creation of the Chemical Abstracts (CA) 9th Collective Index names or in the assigning of the CA Registry numbers. The 9th Collective Index names were implemented to ease the indexing tasks at Chemical Abstracts Service (CAS), and CA Registry numbers are assigned to give a unique identification to each compound without any consideration as to how a compound should be defined. More names are being created, but the primary problems are not being tackled.

As a profession, the chemist has a duty to provide those who are not practising chemists with access to chemical information in the simplest and most informative manner. This can only be achieved by providing chemical name indexes. The profession should take action to ensure efficient access to information on specific compounds. It should discourage the totally misleading practice of searching for chemical names by computer proximity matching techniques. There should at least be sound indexing conventions followed by encoding rules which when used effectively will lead the searcher to pertinent data with a guarantee of accurate retrieval.

PRESENT PROBLEMS IN THE DAY-TO-DAY USE OF NOMENCLATURE

Three major problems with chemical nomenclature at the present time are:

(1) lack of standardisation of nomenclature;
(2) failure to use indexing conventions which will assist the searcher;
(3) the use of unnecessarily large indexes.

The problems that exist and the consequence to the enquirer in his day-to-day use of nomenclature can be identified by examining what is wrong with present methods. The lack of standardisation of nomenclature is considered elsewhere in this book and it is therefore relevant to consider the failure to use conventions and the use of large indexes in this chapter.

The need for improved indexing conventions

The present move to create unique records, as is the case with the CA Registry Number, is only for the benefit of the indexer. Closely related compounds have completely different Registry Numbers as the following example shows:

	Registry Number
Aniline	62–53–5
Aniline hydrochloride	542–11–0
Lactic acid	50–21–5
S Lactic acid	79–33–4
R Lactic acid	10326–41–7

When creating an index, the searcher requires some consideration and the searcher can reasonably expect that various forms of a specific substance are indexed together. The index should be user-related, not indexer-convenient.

Contrary to present practice, devising a coding technique is not the first problem to face, nor is encoding the first step to take when confronted with a compound. Before any translation of a structure into its coded form is carried out, whether the code be a name, a Wiswesser Line Notation or a connection table, it is important that indexing rules are established. These rules should reflect the needs of the searcher. For example, complex salts may need to be coded as the parent; internal salts coded as the hydroxy acid; simple salts need to refer directly to the parent and similar action is necessary with stereo forms. Failure to set up such indexing procedures will inhibit the possibility of establishing relationships within the index. Once the rules for indexing have been established, the next step of encoding can then take place, not with the simple objective of creating a unique record, but a record that is both logically organised and unique.

Creating a logically organised record of a compound

A compound must be considered as a network of the parent and then any modifications or additions to the basic record can be recorded in a sequential manner. If the record of a compound is constructed logically then both registration and substructure searching can be carried out much more easily. A logical ordering scheme for a chemical record could be as follows.

(a) The two-dimensional network of the compound.
(b) Modifications to the network, i.e. substitution of atoms in the parent as in the Na salt.
(c) Additions to the network, e.g. as in the hydrochloride.
(d) Any stereochemical information.
(e) The proportion of constituent molecule present.

A major problem with present indexes is that various closely related forms of molecules are separated. The inverted name index is an attempt to produce a record with the ability to achieve some association but the rules are not rigidly applied. Closely related compounds described in the above way would all appear together in an index.

Present indexing systems expect a considerable knowledge of chemistry even to the extent of whether a particular salt will be formed or whether a stereo form exists. Most searchers of nomenclature indexes do not have this knowledge. Furthermore, even with that knowledge, the searcher must enter an index with a list of parent compound, salts and stereo forms if he is to recover all the information about a compound. If the names were constructed as set out above, then an enquiry would only need to be phrased as the parent structure. Entering an index at that point would give all the information on related compounds in a sequential manner. Also, if the CA Registry Number had been allocated using a similar technique then a useful reference tool would have been established. The present methods of searching using nomenclature or CA Registry Numbers are totally inappropriate for accurate retrieval.

The size of indexes

A further criticism of present indexing practice is the size of the indexes. The number of volumes in the CA 9th Collective Index is indicative of the complication of the present situation. There is no particular advantage in a large and comprehensive name index to most users of name indexes. A chemist does need a comprehensive index, but as stated earlier he does not use nomenclature to enter the indexes. Some of the problems of dealing with over five million components are listed below.

(1) High growth rate.
(2) Large proportion of compounds only of academic interest.
(3) Records are not validated or edited.
(4) Compounds are accessed by chemists using structural methods.
(5) Compounds are accessed by non-chemists via nomenclature, which is tedious and gives inaccurate retrieval.

Non-chemists need access to smaller indexes. The whole of the Western civilisation exists on a handful of compounds, much fewer than 100,000 out of the five million known structures. The advantages of dealing with fewer compounds are:

(1) low growth rate;
(2) data can be validated/edited and related;
(3) indexes can be used by non-chemists by multi-name access: trivial, trade name; etc.;
(4) components can be accurately processed through a structure-based system, e.g. Wiswesser Line Notation or connection table.

Nomenclature is being forced to conform to patterns set up by CAS and nomenclature used at their level bears little relationship to the needs of most users of name indexes. If steps were taken to isolate chemicals into those commonly occurring and those of academic origin, then better name indexes could be provided.

It is necessary to preserve trivial names for many purposes and to index these alongside the systematic names. Cross-referencing, which is rarely adequate in existing indexes, is essential to allow the different categories of users to approach the index from their own viewpoint. Low volume indexes, more fitting to user needs, would allow such indexes to be constructed. The present cesspool approach presented to non-chemist users is inaccurate and inefficient.

SUMMARY

The concept of one name which would fulfil all needs is too great an expectation from nomenclature. The chemist, with his ability to deal with many of his problems in structural terms has abandoned systematic nomenclature to the information profession. But the chemist as a professional person has a duty to the users of chemical information which he is not fulfilling. The chemical profession has played no part in three recent major moves in indexing and searching chemical names, all three of which are disastrous when considered from the viewpoint of accurate retrieval by non-chemists:

(1) the CA Registry Number;
(2) the CA 9th Collective Index name;
(3) computer text searching of nomenclature as offered by the on-line services.

The reason for the chemist not being involved is simply explained. The information scientists have seen the enormous number of compounds as the only problem to be dealt with. The computer offered a method of dealing with this provided certain constraints were accepted, The computer is a logical machine and could have dealt better with the problem if a more logical approach had been taken.

The three main problems in the use of chemical nomenclature can now be identified as:

(1) Nomenclature is expected to play too diverse a role.
(2) Computer methods were adopted ignoring user needs.
(3) The information profession is obsessed with volume.

The final question still remains. If systematic nomenclature is being perpetuated by the information profession then for which users?

CHAPTER 3

The misuse of nomenclature

S. B. WALKER
ICI Plant Protection Division, Bracknell, United Kingdom

One of the best sources of examples illustrating the misuse of chemical nomenclature is reference to catalogues produced by suppliers of chemicals. Every conceivable form of nomenclature will be found in these catalogues ranging from those which claim to use Chemical Abstracts (CA) or the International Union of Pure and Applied Chemistry (IUPAC) nomenclature to those which provide a good meaningful name, and others whose nomenclature leaves a lot to the imagination.

How does this misuse of nomenclature arise? It is probably at school that most would-be chemists receive their first brush with chemical nomenclature, and then as a student, if lucky, some basic training in the way that systematic systems set about naming a molecule will be given. I do not know if any university would consider that more than a day or two at the most should be devoted to a study of the naming of compounds. While some journals are known to request the use of systematic nomenclature, this is difficult to enforce and most chemists would agree that reading experimental data in the published literature does not teach one a great deal. For most chemists, however, it is usually sufficient to be able to give a name to a chemical which is adequate for describing the molecule in conversation with colleagues and in reports. Such a name is sufficient in a local environment where colleagues will be familiar with a particular spectrum of chemistry, but in certain circumstances, for instance for publication, the services of a local expert are required and as often as not, such an expert may not be available. Thus at best, where a systematic name for a molecule is needed it will most probably be a compromise in which a chemical is named in as unambiguous a way as possible. At worst, the chemist will use his limited nomenclature knowledge to suit his own purposes, bending the rules where he finds it necessary, and occasionally making up his own rules as he goes along. There are times when nomenclature alone is apparently not enough, and graphics are required to aid understanding.

Let us try and break down some of the main types of problems encountered.

There are what might be called pseudo-systematic names. Some of these are perfectly valid, although at first consideration one might think that there is something missing from the chemical name, e.g.

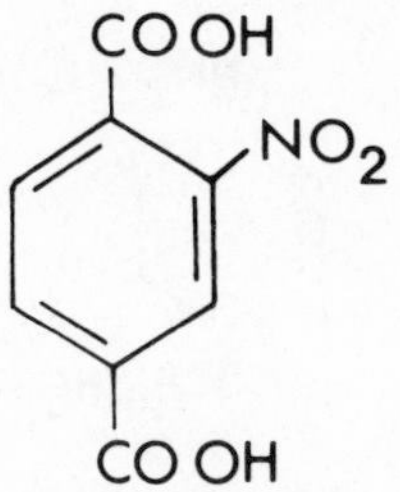

Nitroterephthalic acid

Here the lack of locants for the position of the nitro group must at first appear to be a barrier to deriving an accurate chemical structure. However, because of the relative position of the carboxy groups defined by the use of the word 'terephthalic acid' and the remaining symmetry of the molecule, there is no possible misunderstanding in the positioning of the substituents. However, the name does use a trivial term – terephthalic acid.

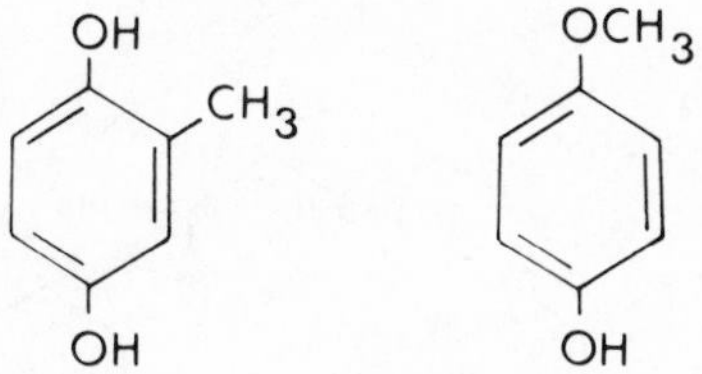

Methylhydroquinone

A similar example that is common to suppliers' catalogues is methyl hydroquinone, again involving a term which gives relative positions to two of the substituents but this time leaving two possibilities for the position of the methyl group which if properly defined could be 2-methyl or O-methyl. The O-methyl has also been misinterpreted as ortho-methyl thus making it equivalent to 2-methyl.

Another catalogue example where a name sounds very scientific, but on examination leaves a number of alternatives, is tetrafluoroethoxy aniline. Here there is nothing to differentiate between the tetrafluoro being ring substituents or substituents which are part of the ethoxy group, and even then there is room for ambiguity as to how the four fluorine atoms are distributed along the ethoxy group. The compound that the suppliers were trying to sell has the middle structure, but of course there is no mention in the name that the amine and tetrafluoroethoxy-group are para- to one another.

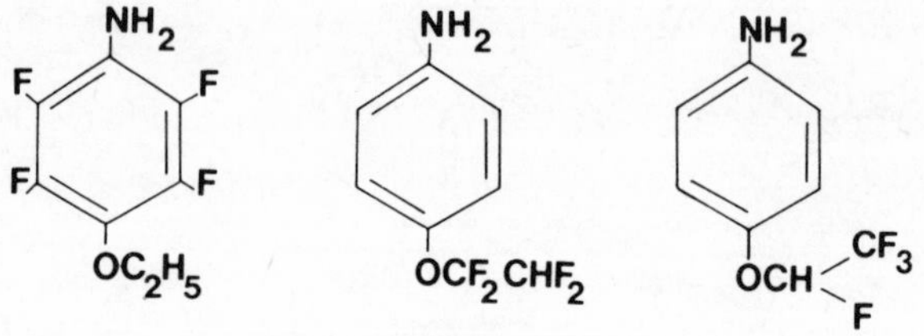

Tetrafluoroethoxy aniline

To derive structures from some chemical names needs local or private knowledge. 3,3′-dibromocresolsulphonphthalein would be immediately recognisable to a dyestuffs chemist who would be familiar with the use of the term 'sulphonthalein' and would know exactly where the 3,3′-positions on the molecule occur.

In work on Crown ethers and related compounds a particular hybrid nomenclature has been developed which enables a short name to be used to replace an otherwise complicated name which would be necessary for these molecules.

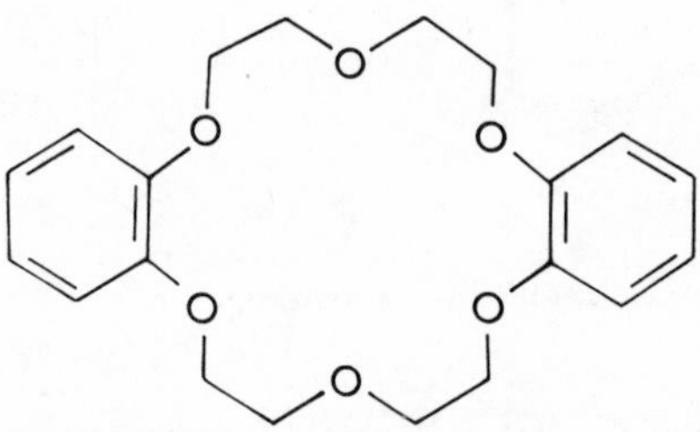

Dibenzo - 18 - Crown - 6

In my own area, the nomenclature of a gibberellin needs specialist knowledge usable in a local environment. How many chemists would know immediately the structure of the gibberellin ring system, and how many would recognise it from the word gibberellic acid? Fewer still would recognise it as (3S,3aS,4S,4aS,6S, 8aR,8bR,11S)-6,11-dihydroxy-3-methyl-12-methylene-2-oxo-4a,6-ethano-3,8b-prop-1-enoperhydroindeno-[1,2-b]furan-4-carboxylic acid.

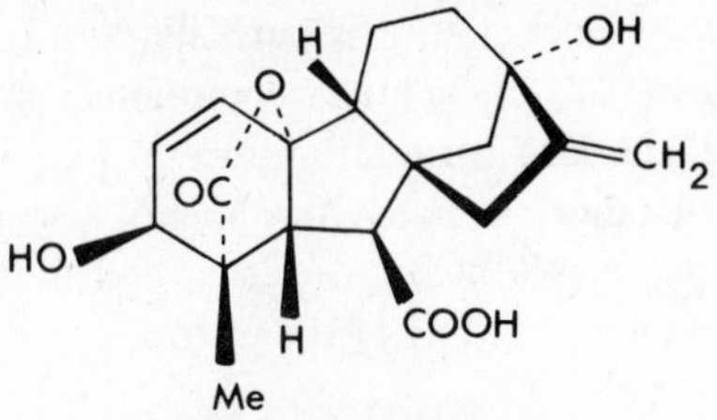

Gibberellic acid

(3S, 3aS, 4S, 4aS, 6S, 8aR, 8bR, 11S) - 6,11 - dihydroxy - 3 - methyl -12 - methylene - 2 - oxo - 4a, 6 - ethano - 3, 8b - prop - 1 - enoperhydroindeno - [1,2-b] furan - 4 - carboxylic acid.

One would certainly need to be party to some very local knowledge in order to draw structures for the following names, recently listed in a journal. The author was no doubt guilty of considerable poetic licence and we should take these names with a pinch of salt (or should I say sodium chloride). They do, however, appear to have a scientific basis.

2,6-Dichlorotolerate
Ethyl decapitate
Copper keystoneate
Polyvinyl fluortile
Hydrofurious acid
Tetrabromoseltzer
Ultrasonic lactone
1,2,3-trimethyl tricycle

Another common source of error is the misuse of a space in a chemical name, for example ethyl-phenyl-malonate, an example of what might be called 'space to taste' nomenclature, an over-generosity of spaces perhaps more to be blamed on a typist or printer than on the chemist originator.

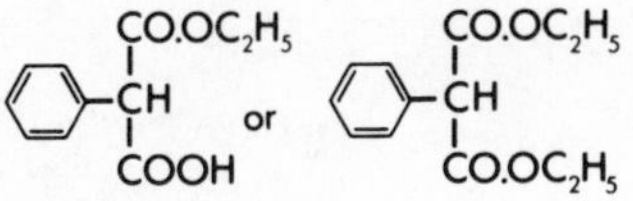

ethyl phenylmalonate

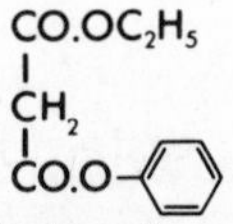

ethyl phenyl malonate

This 'space to taste' nomenclature is particularly a trap for the unwary in nomenclature involving organo-phosphorus compounds and there are occasions where 'bracket to taste' can also form an important part in the correct naming of a compound. For instance, 4-methylthiophenol – where bracketing round (methylthio) shows that the two belong together and that the thio term is not being used in conjunction with the OH of the phenol.

SH / CH_3 — 4-methylthiophenol

OH / SCH_3 — 4-(methyl thio)phenol

Other names that fit in this category involve the mixed use of trivial nomenclature and systematic numbering. Firstly an example in which the use of ortho, meta and para to describe relative positioning of groups on a phenyl ring are taken to a second level.

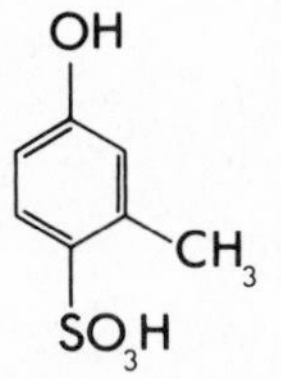

OH / CH_3 / SO_3H

m-cresol-p-sulphonic acid ?

5-Nitro-*o*-toluidine has a different interpretation to a dyestuffs chemist from that which systematic nomenclature would suggest. In one case the 5 is relative to the methyl group whereas most non-dyestuffs chemists would naturally cite the amino-group as the 1 substituent position and place the nitro group relatively at the 5 position.

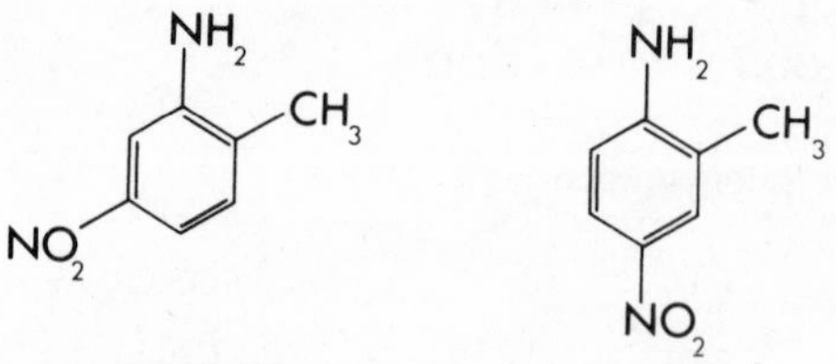

NH_2 / CH_3 / NO_2 — systematic

NH_2 / CH_3 / NO_2 — 'dyestuffs'

5-nitro-o-toluidine

There are quite a few examples of misnomers. Two in regular use are nitroglycerine for what should be correctly termed glycerol trinitrate, and solid hydrogen peroxide and there is even a reference to 'laevo rotatory ice crystals'. There is little more to say about these examples since their use is mainly historical. Everybody understands what nitroglycerine is and understands that it is incorrectly known as such.

There is another class of names which also have their beginnings in history and have survived to the present day. These archaisms are often quaint and usually fairly meaningless unless one is old enough to know their origin. The type of name is tetramuriate of coke for carbon tetrachloride – or more correctly, tetrachloromethane. Chlorinated hydrocarbon molecules must have something about them, because ethylene dichloride is known as 'oil of the Dutch chemist', although its derivation is unknown.

Chemical products may be known by a number of names. There is obviously the chemical name of the active ingredient, but also it will often have a common name approved by the International Organization for Standardization (ISO) in the case of pesticides, or approved by World Health Organization as an international non-proprietary name (INN) in the case of pharmaceutical products. Then there are trade marks and registered trade marks. In coining a common name and sometimes a registered trade mark from a systematic chemical name corruption can take place. For instance, the use of chloroxylenol as an INN – which is no more than a systematic name stripped of the locants which supply the necessary precision – is an attempt to adapt a systematic chemical name which could lead to confusion. Another example is the use of chlorindanol as an approved name for the chemical 7-chlorindan-4-ol.

A reverse situation applies in the name 2,3-dibromotexanol phosphate which shows the treatment of a registered trade mark (texanol) as a systematic parent, which has been adapted, with obvious disadvantages.

There are also some dangers in adapting trivial names. A trivial name is only accepted for a specific compound, but these names can be adulterated by the unscrupulous when trying to name a close derivative. Examples in this category would include 3-chloro-tyrosine or *m*-tyrosine in which the acceptable trivial term for the amino acid, tyrosine, has been further qualified to describe in one case the derivative with a chlorine adjacent to the 4-hydroxy on the phenyl ring, and in the other the moving of the hydroxy group from the para- to the meta-position in the ring.

In coining a common name the aim is to create a name consisting of a few syllables to be used in place of a long and complicated chemical name. Where a perfectly acceptable short chemical name is available, this should be used. An example that comes to mind in the pesticide field here is 3,6-dichloropicolinic acid which has been deemed by the three common name approval organisations, the Americal National Standards Institution, the British Standards Institution and ISO, not to require an alternative common name.

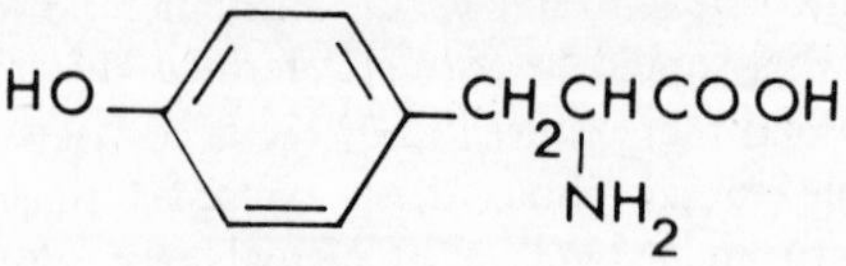

HO–C_6H_4–$CH_2CH(NH_2)COOH$

Tyrosine

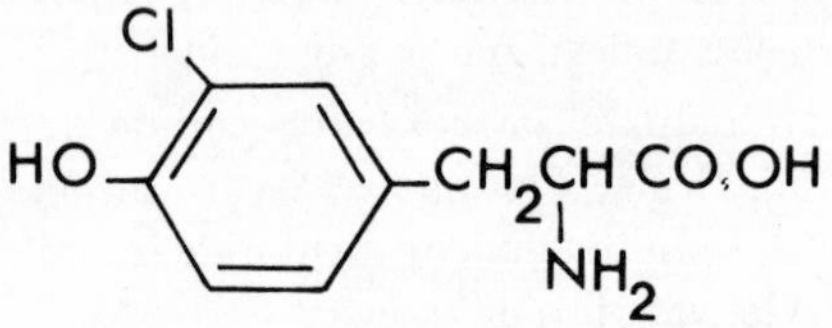

HO–(Cl)C_6H_3–$CH_2CH(NH_2)CO.OH$

3 - Chlorotyrosine

HO–C_6H_4–$CH_2CH(NH_2)COOH$

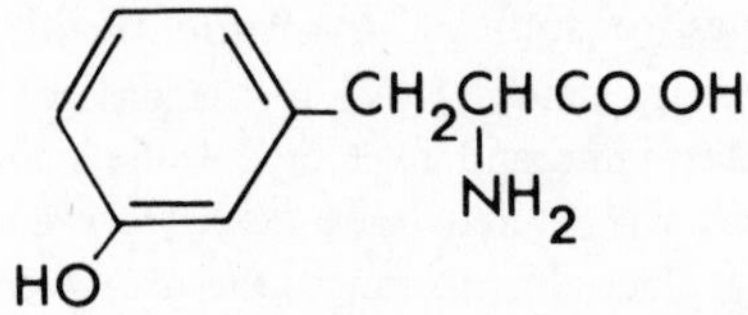

m – Tyrosine

According to the strictest rules of nomenclature, a 'good chemical name' must not only identify a chemical, but the structure must be implicit in the name itself and described unambiguously. And so acronyms and abbreviations are a misuse of nomenclature. Everybody, even a non-chemist, knows of DDT, but would the chemist be able to draw a structure from this name? Acronyms and abbreviations are usually formed from the chemical name. Probably one of the earliest examples of this was EDTA; but while realising that there were quite a number of these acronyms being used in the chemical world, either in catalogues, publications or advertisements, I do not think that I had realised the full extent of their use. Some six months ago, I began to make a note of ones that I found and have compiled these into an index. Although my index is not comprehensive, it has already shown up two serious problems. One is that the same acronym is sometimes used for two quite different chemicals.

DMAP 4-dimethylaminopyridine
DMAP dimethylaminopropylamine

TEA triethanolamine
TEA triethylaluminium

The other is that one chemical may have several acronyms.

DABCO	1,4-diazabicyclo(2,2,2)octane
TED	triethylenediamine
FDNB	1-fluoro-2,4-dinitrobenzene
DNP-F	2,4-dinitrophenyl fluoride

There are some acronyms which are totally unpronounceable and some which produce perfectly valid English words. It is to be hoped, for instance, that the TEA one drinks is not triethanolamine. It may be considered difficult to introduce SEX into a paper on the misuse of nomenclature, but this acronym is commonly used for sodium ethyl xanthate.

In the pesiticide world, the Japanese Ministry for Agriculture and Forestry have for many years used acronyms as common names, e.g. MIPC for methyl isopropyl phenylcarbamate, DAPA for 4-dimethylaminophenyldiazosulfonate.

Acronyms and abbreviations are fashionable in other sciences and professions as well as chemistry, but neither can ever fulfil the requirements of a systematic name. The proliferation of acronyms can only lead to further duplication of the type mentioned, where two compounds can share an acronym, and one compound could have several.

Although no obvious acronyms exist for CABS or IUPAC, these could be:

$(CH_3)_2CHNHCONH$–(benzene ring)–CH_2COOH

IUPAC

4-(3-isopropylureido)phenylacetic acid

(cyclohexane ring)–$NHCH_2CH_2CH_2SO_3H$

CABS

4-(cyclohexylamino)butanesulphonic acid

These show the way the letters forming the acronym are abstracted from the chemical name.

When I first started in my present job I discovered that there was a room on site in which chemicals, ordered by chemists but not used up, were stored.

The chemicals in this store were indexed by supplier's name and on the wall were 'Instructions for use of the Store'. These read:

1. Think up a name for the chemical that you need.
2. Look this name up in the index.

   If you find the chemical, note its location and sign out the chemical on the card.
3. If you don't find the chemical – think up another name, and look the new name up in the index.

These instructions conjured up visions of solitary chemists sitting in the store, industriously inventing names for a desired chemical and going grey in the process.

Most chemists of course do not search a store cupboard for their intermediates. When they require a specific substance to use as an intermediate in the preparation of an agrochemical or pharmaceutical product, how can they go about looking for the material? They are faced with an array of catalogues which may contain the chemical required. They may choose in the first instance, given that, as already indicated, most chemists are not experts in nomenclature, to work out the molecular formula for the chemical and in some chemical catalogues they will be helped because the suppliers have taken the trouble to produce a molecular formula index. This has an obvious advantage in that they have not had to think about a chemical name. But this will only allow a search of perhaps half a dozen catalogues, and if this approach fails they have to tackle the naming problem, The immediate reaction would be to name it along the lines of their synthesis thoughts, but of course this may not be the name used by the supplier, Even though the compound sought by the chemist may well be present in a number of catalogues, it will almost certainly be present under various, totally different, names. It may even be present in the same catalogue under totally different names!

The problem of locating a variously named compound can be successfully overcome by the use of the Wiswesser Line Notation (WLN). An index has been produced which is based on the fact that for any specific molecule there is only one WLN whereas there may be many systematic, trivial and common names.

However, the need to produce such an index emphasises the problems of naming chemicals for sale. My experience in the production of this index highlighted some aspects not obvious to the casual reader of chemical suppliers' catalogues. For instance, there are numerous examples of a supplier selling the same chemical under quite different chemical names and at very different prices depending on which name is selected. One supplier was found to have in its catalogue:

2-naphthol
β-naphthol
and 2-hydroxynaphthalene

By buying the material under the right name, the chemical could cost one-tenth of the price listed for one of the alternative names and, in this case, no indication of a difference in purity. The best example found involved a price differential of twenty-six times for two totally different chemical names for the same material.

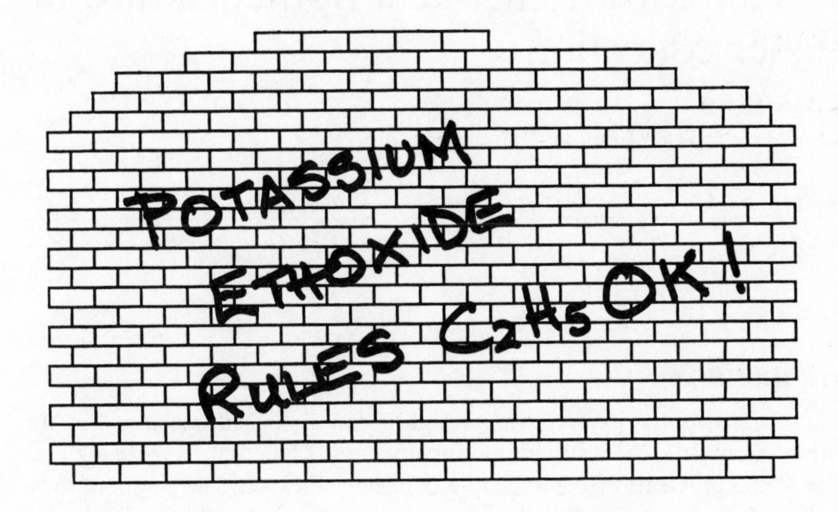

The study of clever and witty graffiti is much appreciated these days by those intelligent enough to understand it. The key to understanding chemical nomenclature lies in better education. Misuse is directly attributable to a lack of proper training.

CHAPTER 4

Problems with chemical nomenclature in higher education

Dr G. P. MOSS
Queen Mary College, London, United Kingdom

INTRODUCTION

While claiming some expertise in organic chemical nomenclature and in the teaching of organic chemistry at a university level, I have only a limited knowledge of the problems of nomenclature in higher education. The questions to be considered in this chapter are therefore largely concerned with communication in the broadest sense of the word.

There are many attitudes adopted by students, and staff, to nomenclature. Figure 1 shows one response to this problem. Although a cartoon illustrates a typical response, 'why change the name everyone knows?', fortunately most of my colleagues in education take the alternative view that students should be familiar with a range of nomenclature systems and even encourage the standardisation which the International Union of Pure and Applied Chemistry (IUPAC) endeavours to achieve. This problem can perhaps be illustrated in another way using a quotation from Romeo and Juliet:

> What's in a name? That which we call a rose
> By any other name would smell as sweet.

Various names for a single compound are given in Figure 2. The one that most chemists would best recognise is the graphic formula – not the four names which are the concern of nomenclature committees.

HANDWRITING

In higher education we have to cope with all types of names but since the main methods of communication are oral, backed up with handwritten material on a blackboard, there are several problems not encountered with published material.

"What idiot changed the label on this bottle?"

Figure 1
(Reproduced from *Trends in Biochem. Sci.*, 1980, **5**, 1, by permission of Elsevier Biomedical Press.)

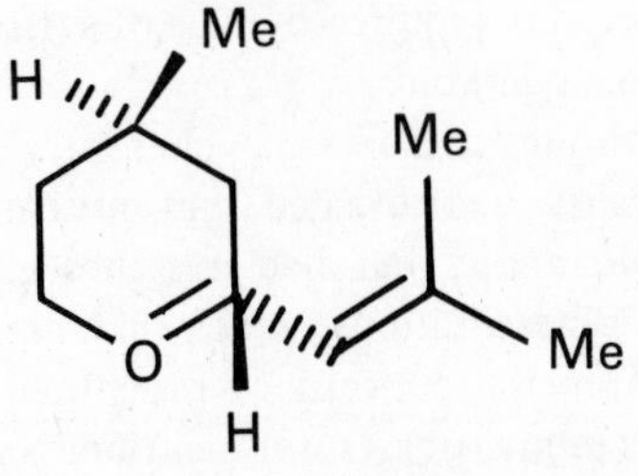

trans-Rose oxide
(2*R*, *trans*)-tetrahydro-4-methyl-2-(2-methyl-1-propenyl)-2*H*-pyran
(2*R*.4*R*)-tetrahydro-4-methyl-2-(2-methylprop-1-enyl)pyran
erythro-4,8-epoxy-2,6-dimethyloct-2-ene

Figure 2

When handwritten, many letters are difficult to distinguish. Table 1 shows some of the confused letters, problems not only encountered by students but also by a typist dealing with handwritten material.

Table 1

Roman/Greek	n η u μ v ν a α B β p ρ T τ X χ w ω (also U V in Roman text)
Lower/Upper case	c C k K o O p P s S u U v V w W x X y Y z Z
Letter/number	O 0 l 1 Z 2 S 5

AIDS FOR TEACHING NOMENCLATURE

The second edition of the ASE document on chemical names for schools has been mentioned earlier (Chapter 1(ii)). Its use has given most students a good basic grounding in systematic nomenclature. However, at college there is sometimes a lamentable ignorance of trivial names such as chloroform, acetic acid or acetone. In order to find out more about any problems encountered in higher education I made enquiries on the teaching of nomenclature rules, covering most chemistry departments in the United Kingdom. To generalise, most departments rely on covering nomenclature in lectures as and when required. This is backed up by textbooks, duplicated notes and tutorials, though the subject only forms a minor part of the curriculum. There are a few aids available; of particular note is a good introductory booklet by Dr Niel S. Issacs of the Chemistry Department, Reading University, *Naming Organic Compounds.* Another aid, produced by Wolverhampton Polytechnic, is a cassette on this topic with accompanying notes.

TEXTBOOKS

Unfortunately it is inevitable that most students rely on textbooks for further details to amplify their lecture notes. Almost all relevant textbooks are American and the style and spelling they see in these textbooks is American rather than English. Another problem is that organic chemistry is still largely taught via functional groups. Hence this requires the use of suffixes but rarely the prefixes needed for polyfunctional compounds.

One of my special interests is in stereochemistry. A check on a number of textbooks in this specialist area revealed that most mention the use of *R* and *S,* and of *E* and *Z,* but about half had mistakes in the assignment of these symbols through over-generalisations which in complex cases may give the wrong result. Furthermore the concept of prochirality, of great importance in bio-organic chemistry, is in many cases not mentioned at all.

ORAL COMMUNICATION

Another problem in the usage of nomenclature is that of pronunciation. Here the concern is less with examples such as ethyl (e′thil or eth′il) or amide (a′mid or am′id) than with the very similar endings such as -ane, -ene, -yne, -one which are easily confused without care. Similarly -ol and -al are readily confused not only in speech but also in handwritten material. An example from a University of London external practical examination illustrates this problem. The unknown set was *o*-chloro-acetophenone but over the telephone the laboratory superintendent interpreted the substance as ω-chloro-acetophenone. The mistake was rapidly found out in the examination as the latter is a strong lachrymator so that all present were rapidly reduced to tears. Another illustration concerns a user with a colonial accent who ordered a cylinder of ethane and was surprised to be sent acetylene (ethyne).

A related problem may occur with metathesis when syllables are interchanged. For example alanine could be converted into aniline which might have disastrous consequences for the student.

ALTERNATIVE NAMES

Nomenclature has evolved over the years and is, of course, still changing. Hence it is not adequate in higher education to deal just with the present state of the art, but also to cover older systems. By the third year at Queen Mary College students are expected to have some idea of the different names used by Chemical Abstracts (CA) and to emphasise this point an exercise is set to find the names used for a compound in each of the decennial (quinquennial) periods. This helps to counter the claim that the chemistry of compound 'X' stops after, say, 1956, whereas all that has happened is that the name has changed. The problem is even worse when teaching the history of chemistry. It may perhaps be still relevant to deal with names such as muriate of potash. However, when it is realised that the chemistry has not changed, only the names, then a general knowledge of chemistry is often enough to interpret the names.

Transatlantic differences only occasionally cause much confusion with students. Some of the more obvious examples are listed in Table 2. It is the last three compounds in this list which are most likely to cause problems for users, for example, in the use of indexes. The spelling of sulphur is nearly a lost cause. With American textbooks naturally using the more logical sulfur it is almost impossible to keep correcting 'f' to 'ph' in written work. While there is no justification for the loss of the 'i' in aluminium there is perhaps a case for the loss of the 'a' from caesium. There are promising signs that the more logical British style for the location of the locant will become more widely adopted.

The value of alternative unambiguous names has already been emphasised in earlier chapters. In research, the use of appropriate names may readily show the relationship between compounds in a way that, for example, the CA index name would not. This has always been so and may have many advantages. The first master of the chemistry department at Queen Mary College (then the Peoples' Palace Technical School) worked on the atomic weight of gold. The obvious source of gold was the Royal Mint sited near to the School. Regulations meant that gold could only leave the mint as coins. However, the administrators of the Royal Mint were quite happy for him to borrow some sodium aurate.

Table 2

UK	US
sulphur	sulfur
aluminium	aluminum
caesium	cesium
butan-1-ol	1-butanol
but-2-en-1-ol	2-buten-1-ol
adrenaline	epinephrine
haem	heme
oestrane	estrane

A further example relates to the early days of the polymer industry when vinyl cyanide was used as a monomer. When the process was scaled up for industrial production the interstate authorities stepped in to forbid the shipping of this 'dangerous cyanide'. However, they were quite happy to permit the passage of acrylonitrile.

CAMERA-READY COPY

In research the ultimate objective of the work is to publish the new results. These days there is increasing use of camera-ready copy and this in turn places much greater responsibility on the author. An area where great care is required is in the drawing of formulae. This is perhaps well illustrated by an example which went with the statement that 'formulae that display stereochemistry should be prepared with extra care so as to be unambiguous'. The author's submission is shown in formula (1) of Figure 3. This appears satisfactory until the reader

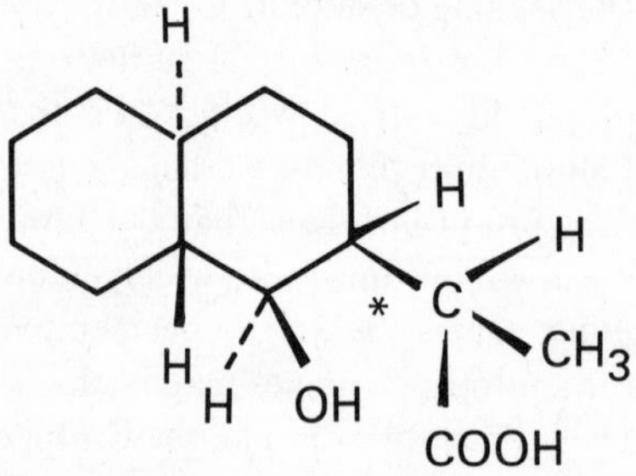

H H H H * C CH3 H H OH COOH

(1)

R H C COOH CH3

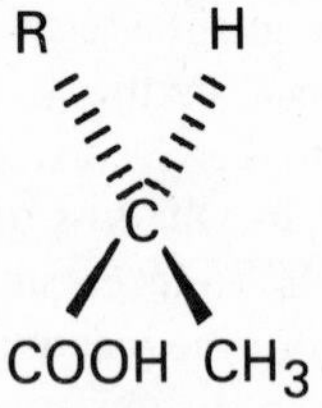

(2)

Figure 3

closely inspects the chiral centre marked with an asterisk. This may be redrawn as shown in (2) which is another way of showing a planar carbon atom. Hence the very purpose of the diagram is in fact lost.

Care is also needed with nomenclature. Problems have arisen, for example, through the use of incorrect names for permethrin and fosetyl-aluminium on first publication. Even more difficulties arise when names are translated by newspapers. An example which appeared in *The Times* transformed adenosin-2′,3′-phosphate into adenosin-2ft,3ft-phosphate.

Another source of confusion may arise with the use of abbreviations. The importance of always defining abbreviations can be illustrated by an example in a nutritional paper where the Krebs tricarboxylic acid cycle was labelled on a diagram as the TCA cycle. Unfortunately the editor expanded the abbreviation which appeared as the trichloroacetic acid cycle. One of the most common problems arises with the abbreviation Bz. In many contexts it is not clear whether benzyl or benzoyl is implied. It is just as easy with only an odd additional letter to use $PhCH_2$ or PhCO which are always unambiguous.

The increasing use of camera-ready copy has placed a greater importance on the use of the correct printing conventions. The significance of round or square brackets with an isotopically substituted or labelled compound, or to indicate a side chain or repeating unit in the main chain, is something which is often left for editors to deal with. It is important that authors use the 'Blue Book'.[1]

FEEDBACK

Feedback from users is necessary information for nomenclature committees. Whilst such committees may not be able to act on all proposals, members consider all proposals and welcome help from users of the 'Blue Book'. In turn I make a plea that CA make their notations clearer to the user. Two recent examples indicate this problem. Dieldrin (Figure 4) gave the wrong stereochemistry using the CA name. The conventions used by the Chemical Abstracts

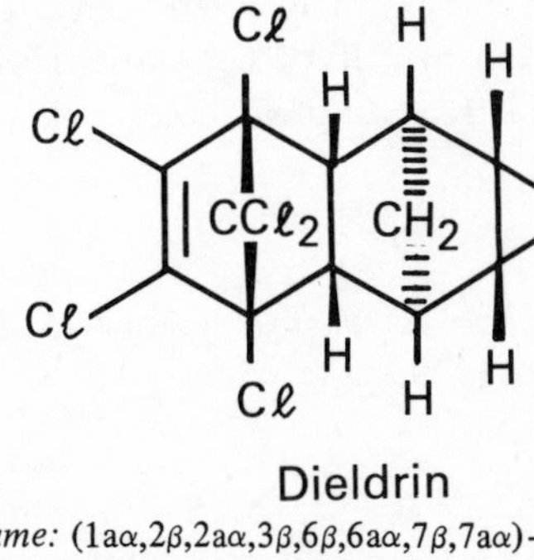

Dieldrin

Chemical Abstracts name: (1aα,2β,2aα,3β,6β,6aα,7β,7aα)-3,4,5,6,9,9-hexachloro-1a,2,2a,3,6,6a,7,7a-octahydro-2,7:3,6-dimethanonaphth[2,3-*b*]oxirene.

IUPAC name: (1*R*,4*S*,4a*S*,5*R*,6*R*,7*S*,8*S*,8a*R*)-1,2,3,4,10,10-hexachloro-6,7-epoxy-1,4,4a,5,6,7,8,8a-octahydro-1,4:5,8-dimethanonaphthalene.

Figure 4

Service to decide priority, and hence α or β, are not adequately explained in the *Index Guide.* At least the IUPAC name is unambiguous.

Frequently there is added confusion with the drawing of this type of compound where the bridgehead substituents are given the opposite stereochemistry to the bridging group. This results in an apparent planar carbon atom.

The last example comes from a well-known drug firm who were confused by 11-HETE (Figure 5) an important fatty acid related to the prostaglandins and leucotrienes. Here the problem arises because the CA name does not indicate the stereochemistry in locant order but in priority order, a concept which is only explained in the least widely read paper on the use of *E* and *Z*, and not repeated in the *Index Guide* or the *Journal of the American Chemical Society* version of the paper.

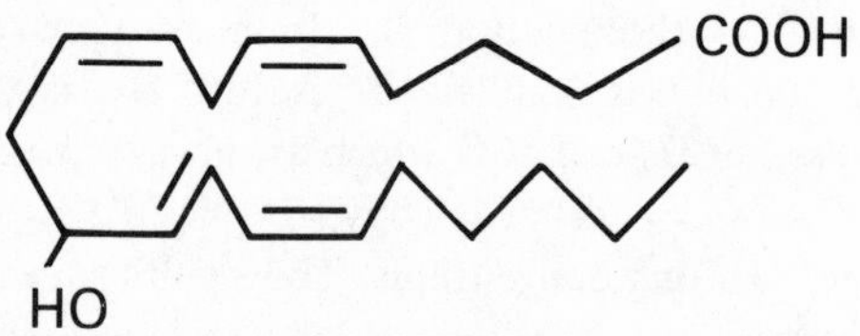

COOH

HO

11–HETE

Chemical Abstracts name: (*E, Z, Z, Z*)-11-hydroxyeicosa-5,8,12,14-tetraenoic acid.
IUPAC name: (5*Z*,8*Z*,12*E*,14*Z*)-11-hydroxyeicosa-5,8,12,14-tetraenoic acid.

Figure 5

CONCLUSION

The need for training in organic nomenclature at the higher education level requires a good working knowledge of current practices, as well as past systems. With the use of camera-ready copy there is a greater importance for authors to apply conventions which in the past have often been dealt with at the editorial stage. The universal adoption of IUPAC conventions with the elimination of national variations would greatly assist this process.

REFERENCE

1. *Nomenclature of Organic Chemistry,* sections A, B, C, D, E, F and H, Pergamon Press, Oxford, 1979.

CHAPTER 5

Standardisation of chemical nomenclature

Dr K. L. LOENING
Chemical Abstracts Service, American Chemical Society, Columbus, Ohio, USA

Some time ago I had a British secretary. Soon after she came to work for me I left for a meeting overseas. During my absence she received a telephone call for me. 'Oh', she said, 'I'm terribly sorry, but Dr Loening has gone to the United Kingdom.' There was a shocked silence for a moment and a hushed voice said, 'My goodness, I didn't know – am I too late to send flowers?'. May be this story says something about the problems of communication that users of chemical nomenclature face from time to time.

The need for standardisation in chemical nomenclature may appear obvious if the term is defined too narrowly. However, chemical nomenclature is the total language of chemistry and relates to everything that people talk or write about when conveying chemical information. It covers a much wider field than simply assigning names to chemical compounds; it embraces quantities and units, symbols, acronyms, abbreviations and even pronunciation. Regarded in this way, nomenclature standardisation needs re-examination. Three basic levels of usage of chemical nomenclature can be considered. Firstly a very informal level, used in oral communication between people who know and understand each other. This level can include nicknames or laboratory codes and is unstructured. Secondly a more formal level used in written documents, scientific papers, reports, and so on where the writer has to be sure when addressing a large readership that every reader will understand what has been written. The writer has to be precise, accurate, unambiguous, but can still retain flexibility in the types of name he chooses. It may not always be the case that systematic chemical names are the best for conveying his meaning. An example of this level of usage taken from my own researches concerns the esterification of some hindered aliphatic acids. Acetic acid was the chosen standard and the work related to steric hindrance. I chose the name methylacetic acid when measuring the esterification rate of propanoic (or propionic) acid, because this conveyed the idea that a methyl group had replaced the hydrogen in the standard material, i.e. acetic acid. Thirdly, there is the exceedingly rigid and precise level of nomenclature that is required for compendia or indexes such as those published by the Chemical Abstracts Service (CAS), where the indexer, and the user, have to arrive at a single preferred name every time.

The language of chemistry or chemical nomenclature is not only used by chemists. All kinds of scientists – biologists, people in the health profession, environmentalists, physicists, engineeers – have a need for a chemical language. That this need extends to book publishers, journalists and the legal profession can be illustrated by two examples. In 1964 a banker in the United States asked for my help on a problem relating to chemical nomenclature. The request concerned the symbols for water and for gold. Several weeks later bumper stickers started to appear on cars with the message 'Au–H_2O: Goldwater for President!'. Here was an interface of chemical nomenclature with politics. The second example concerns a professor at a university in the eastern states who explained that a celebration was being organised for the 60th birthday of his colleague Professor Robert B. Woodward. The composer Leonard Bernstein was commissioned to compose a piece of music for the occasion and he suggested that some of the natural products which Woodward was the first to synthesise should be incorporated in the score. He therefore sought chemical advice on the long systematic names for some of these products, names he felt would be more suitable to compose music to than the short everyday names. I was able to supply the systematic nomenclature for vitamin B_{12}, cortisone, and other relevant compounds. Although the project unfortunately did not come to fruition owing to other commitments, the author of this contribution can truly claim to be a lyricist to Leonard Bernstein.

Chemical nomenclature is not static; it changes with time just as the science of chemistry changes. For example, Arrhenius and Bronsted and Lewis and others have changed our conception of acids over the years. In certain circumstances it is now advisable to specify whether the reference is to a classical acid or a Lewis acid. Again, the meaning of the word 'radical' has evolved during this century. In the early years of chemistry a radical meant any kind of group; for example, 'methyl radical' could be used in place of 'methyl group'. A modern chemist uses 'radical' to mean a free radical, while a radical in the old sense is described as a methyl group or methyl substituent or methyl analogue or whatever the appropriate case may be. So the meaning of old terms in the language of the chemist slowly changes.

In addition, new terms are needed as chemistry expands into new areas. New chemical compounds are being synthesised and new ideas concerning them are put forward. Many chemists would like to coin distinctive names for substances they have made, but they should be cautioned against doing this thoughtlessly. It is most unwise to pollute the chemical literature with a new term when it is not required. If it is really necessary, the rule should be to avoid the use of an old name with a new meaning. This principle is sound and should be observed. It is essential that the new term should not be confused with one already in use. The second useful rule is to derive the term from Latin or Greek. Too many chemists give little thought to the international consequences of their newly coined terms. Although more and more chemistry is being written in English there are some areas, for example cage systems, where trivial names abound because the systematic names are very long and unwieldy. Nomenclature such as 'snoutene', so named because the graphic formula looks like a snout, is lost on chemists whose native language is not English. If this kind of descriptive name must be used it should be translated into classical Latin or Greek, thereby increasing the chance of its being universally understood. In 1951 Drs Crane and Patterson of the CAS fought against the newly introduced name petrochemicals. They maintained, quite correctly, that petrochemicals and petrochemistry should be respectively the chemicals and chemistry derived from rocks rather than petroleum. CAS lost this fight, but fortunately this particular new term did not create as much of a problem as had been feared. However, the principle set out above is a sound one and should be followed.

The current Chemical Abstracts (CA) index nomenclature arouses strong feelings among chemists. It is praised to the sky by some and damned by others. But it should at least be noted that the many changes made by CA in 1972 did not include a new vocabulary. CA eliminated certain trivial names (for example 'isopropyl' was replaced by '1-methylethyl'), but chemists did not have to learn anything new; the meaning of methyl and ethyl were well established. Like CA, authors of chemical papers should exercise restraint in introducing new nomenclature. Although there are official bodies which approve or reject new chemical terms, there is still quite a time-lag. If a paper relates to an area in which many research papers are being written within a short period and these papers are passed from group to group, undesirable names can establish themselves. By the time an official committee is able to disapprove of the new nomenclature, the literature already contains many examples.

I do not think it necessary to justify the need for at least some standardisation. This need can be illustrated by the following examples. In World War II when I was in the US Chemical Warfare Service, we were concerned with the use of phosphorus compounds, but at that time there were no standards for naming them. On one side of the Atlantic, users would talk, for example, about chlorophosphonates in which esters or salts of phosphonic acid, which has a phosphorus–hydrogen bond, had the hydrogen replaced by chlorine. On the other side of the Atlantic the same compounds were called chlorophosphates, based on the old inorganic nomenclature where a hydroxyl group has been replaced by chlorine. Government and military personnel had enormous difficulties in communicating on these compounds because of the different nomenclature systems. Again, in the early 1950s there was great interest in the USA about potential rocket fuels, and this interest led to a considerable increase in work on boron chemistry. Most of the work was classified and both the Air Force and the Navy had an enormous number of projects and contractors. Those engaged in the work had difficulty understanding one another's reports, because there was no agreed system of boron nomenclature and every laboratory expressed things differently. So, at the instigation of the US Air Force, the first meeting of the American Chemical Society (ACS) on boron nomenclature was held in 1954.

The means by which chemical nomenclature is standardised can be considered on two levels. Firstly in the USA and secondly internationally. At a national level, the ACS is divided into various divisions. Many of these, for example the Divisions of Organic Chemistry, Inorganic Chemistry, Carbohydrate Chemistry and Physical Chemistry, have their own nomenclature committees which make proposals that eventually ascend to the international level. The process also works in the reverse direction and provides a forum for international proposals to be examined by people working in the field who have a special interest in any decisions that are reached. An individual chemist has an opportunity through these committees to offer comments and criticisms. The divisional committees are overseen and represented by a general committee, the ACS Committee on Nomenclature, the main role of which is to ensure that the various sub-disciplines do not go their separate ways and create new conflicts and misunderstandings. Internationally there is a similar structure to that in the USA. The International Union of Pure and Applied Chemistry (IUPAC), like the ACS, is organised into divisions such as Physical, Organic, Inorganic, Macromolecular, Applied and Clinical Chemistry. Each division has a commission on nomenclature and it is within these that the actual work is done. Recommendations from each of the Nomenclature Commissions over the last five years are approved by each division and then sent to the Interdivisional Committee on Nomenclature and Symbols (IDCNS). This Committee fulfils a vital function in co-ordinating proposals and avoiding conflicts, and is the international counterpart of the ACS Committee on Nomenclature. Prior to the formation of the IDCNS there were occasions when IUPAC recommendations from divisions conflicted with other IUPAC recommendations. IDCNS does not operate as just another layer of bureaucracy; instead, its function is to ensure that potential conflicts are avoided. The IDCNS handles conflicts not merely within IUPAC itself but also between its recommendations and those of sister unions such as the International Union of Biochemistry, the International Union of Nutritional Sciences, the International Union of Pure and Applied Physics, the various committees of the International Organisation for Standardisation, and so on. IDCNS is the authority for IUPAC in matters of nomenclature and IUPAC publishes nomenclature recommendations in books, colour-coded according to subject. Green is used for physicochemical nomenclature, red for inorganic, blue for organic, and orange for analytical.

Three problems still remain which inevitably face users of nomenclature. Firstly, it is important to recognise that there will always be a time-lag between the introduction of new terms in a fast-growing area and the issue of official recommendations concerning them; efforts must be made to reduce that lag as much as possible. Secondly, it is essential for the benefit of users that nomenclature recommendations should be sound and well-based. The views of experts working in the various fields are essential in developing these recommendations. Without their input the recommendations will be defective and liable to be disregarded. A particular problem is that work in nomenclature is done mostly by veterans; contributions from younger chemists are few. This fact can be attributed to the lack of recognition given to this type of work. A young academic can use his time to bring in a research grant or a teaching award to help his professional career, but service on a nomenclature committee is unlikely to assist him professionally. So some way must be found to provide the kind of recognition for nomenclature work that would encourage younger chemists to join the committees. A similar problem exists regarding industrial involvement. Many of the nomenclature bodies of IUPAC could make use of more members drawn from industry as well as increased recognition and support from the world of commerce. Additional financial support is also necessary because IUPAC, like everybody else, has problems raising sufficient money in these inflationary times.

The third problem particularly concerns educationalists. It is important for a teacher to care about the language of the discipline in which he is working. Entirely apart from the provision of specific courses in nomenclature, emphasis must be placed on the writing of good English and good chemistry also, with attention paid to the principles on which our chemical language is based. Furthermore, these principles should be instilled throughout the educational process.

CHAPTER 6

Standardisation of chemical nomenclature in Government legislation and documentation

E. W. GODLY
Laboratory of the Government Chemist, London, United Kingdom

The terms used in this title require some preliminary definition.

Government legislation and documentation. At a national level there are Acts of Parliament such as the Finance Act, Government White Papers, Regulations, Information Booklets, etc., as well as such documents as British Standard (BS) Specifications to which legislative instruments may make reference. On an international level there are in the European Community (EC), Regulations and Directives, for example the EC Directive on the labelling of dangerous substances, and for world-wide use such international conventions as the Nomenclature of International Trade Goods of the Customs Co-operation Council, which includes chemicals.

Documentation can cover chemicals from a variety of aspects of interest, such as safety and hazards in storage and transport, environmental contamination, food and drug legislation, labelling, tariff classification, and also such non-statutory matters as production and trade statistics and patterns of consumer behaviour.

Chemical nomenclature. This may mean all the ways in which identifying information may be conveyed not only for chemicals but also reactions, methods, and even apparatus. In this chapter it is used only for the naming of chemicals in words, i.e. the assignment of chemical names. Chemical names are analogous to names of people, but should be completely distinctive as shown in Figure 1. The information in the left-hand column of Figure 1 may suffice to identify the individual, but they are not names; the right-hand column entries are the relevant names. Thus, a chemical name is a formal entity and it can, with the usual conventions, take its place in an alphabetical list.

Identifying description	Name
The fourth son of a Liverpool merchant who became Prime Minister	William Ewart Gladstone
Benzoic acid, 2,4-dinitro derivative, sodium salt	Sodium 2,4-dinitrobenzoate

Figure 1

Standardisation. This is a theoretical concept by which each chemical would be known by one name only in all contexts. Theoretical because, apart from most of the elements and a few simple compounds such as water, each chemical has several names in English alone. Nevertheless, there is pressure for such standardisation. There are about 15,000 chemicals of trade importance, 50,000 of interest in environmental work and about seven million altogether. The large number of synonyms raise particular difficulties for the user. They magnify the work of librarians, indexers, computer-people, traders, transporters, governments and their servants, statisticians, students, teachers, in fact everybody. As far as legislators are concerned, they seldom feel very strongly on the choice of the name, provided the chemical nomenclature can be understood and does not get changed once it has been used.

Unfortunately, chemical nomenclature, like all linguistic communication, develops continuously, and the ideal of 'one name per chemical' is not to be realised without great efforts of organising ability on the part of nomenclature specialists and evangelical zeal in the propagating of their recommendations. It will also need open-minded progressiveness on the part of those called upon in the common interest to discontinue the use of well established names.

Chemical names may be coined by discoverers of new substances either with or without some attempt at a system, but which system? Various schemes have been suggested and a few actually applied. Specialised systems for special fields have been developed alongside general ones. The names in the *Colour Index* are doubtless well understood in the dyestuffs industry, but probably not outside it. For example, how many non-dyestuffs chemists know the identity of Fast Scarlet Salt R?

Figure 2 illustrates the variety of names available for one chemical. The thirteen names are all more or less systematic and all give the correct structure. None of them break the rules of chemical literacy with wrong punctuation or violation of the alphabetical order of prefixes principle. To do so would generate even more names. Even the simple butan-1-ol can also be 1-butanol or butyl alcohol.

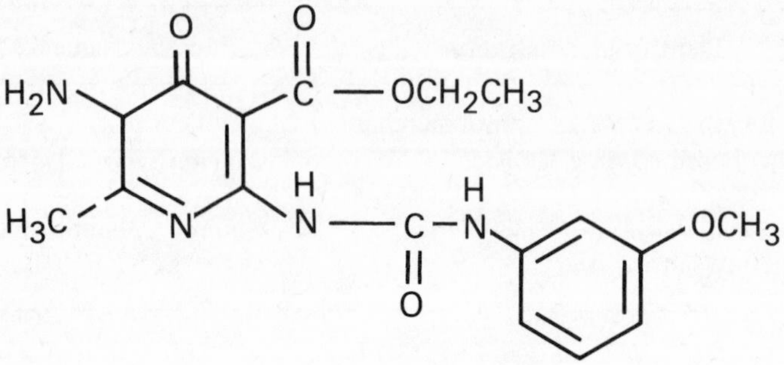

Correct application of the IUPAC rules gives the following name for the above structure:

Ethyl (5*R*)-5-amino-4,5-dihydro-2-[3-(3-methoxyphenyl) ureido]-
5-methyl-4-oxonicotinate

If stereochemical considerations are neglected, the following attempts, although giving the correct structure, must all be rejected for one reason or another. Some are almost respectable; a few are rather far-fetched. Further efforts might also give the right structural formula but would be chemically illiterate.

1 Ethyl 5-amino-2-[3-(3-methoxyphenyl) ureido]-5-methyl-4-oxopyridine-3-carboxylate
2 3-[3-(5-Amino-3-ethoxycarbonyl-5-methyl-4-oxopyridyl) ureido] anisole
3 1-(5-Amino-3-ethoxycarbonyl-5-methyl-4-oxo-2-pyridyl)-3-(3-methoxyphenyl) urea
4 3-[3-(5-Amino-3-ethoxycarbonyl-5-methyl-2-pyridyl)ureido]phenylmethyl ether
5 6-(*m*-Anisidinocarbonylamino)-5-ethoxycarbonyl-3-methyl-4-oxo-3-pyridylamine
6 *N*-(5-Amino-3-ethoxycarbonyl-5-methyl-4-oxo-2-pyridylcarbamoyl)-*m*-anisidine
7 3-Methoxyanilino-*N*-(5-amino-3-ethoxycarbonyl-5-methyl-4-oxopyridine)-
2-formamide
8 5-Amino-3-ethoxycarbonyl-2-(3-*m*-methoxyphenylureido)-5-methyl-4-pyridone
9 Ethyl 5-amino-2-[3-(3-methoxyphenyl) ureido]-5-methyl-4-pyridone-3-carboxylate
10 Ethyl 5-amino-*N*-*m*-methoxyphenylcarbamoyl-6-oxo-5-methyl-3-aza-anthranilate
11 5-Amino-3-ethoxycarbonyl-2-[3-(3-methoxyphenyl) ureido]-5-methyl-4-oxopyridine
12 Ethyl 3-amino-6-[3-(3-methoxyphenyl) ureido]-4-oxo-3-picoline-5-carboxylate

Figure 2

Faced with this scenario, the Chemical Abstracts Service (CAS) has devised – not without a few convulsions – a set of rules for reducing this variety down to one name per chemical and published the *Index Guide (1972–6)*[1] to show roughly how this can be achieved. Why then do Governments not simply use the CAS name when one is required? The answer is three-fold. (a) To do so would often mean discarding established familiar names such as aniline and isobutane in favour of the newer and stranger benzenamine and the heavier 2-methylpropane. (b) The CAS name is often more cumbersome than that in current use the world over and often comes with many seemingly superfluous brackets. (c) A more fundamental objection is that CAS names often consist of the parent, listed alphabetically, followed by qualifying phrases conveying particular derivatives.

Unless the searcher is trained in the CAS rules, this parent is not always easy to identify for a given structure. Figure 3 illustrates three examples of this type

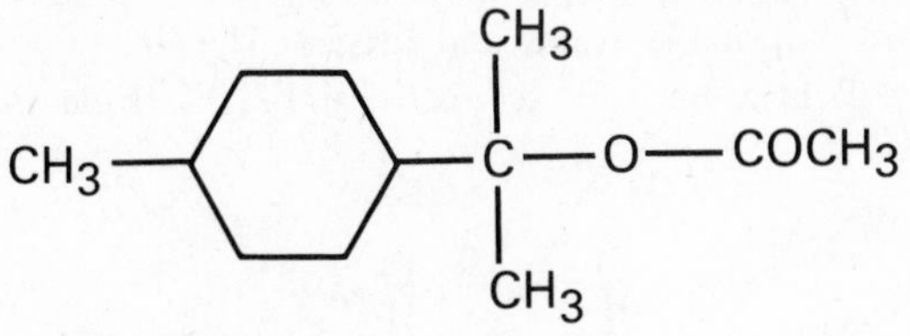

1-methyl-1-(4-methylcyclohexyl)ethyl acetate [IUPAC]
cyclohexanemethanol,α,α,4-trimethyl, acetate ester [CAS]

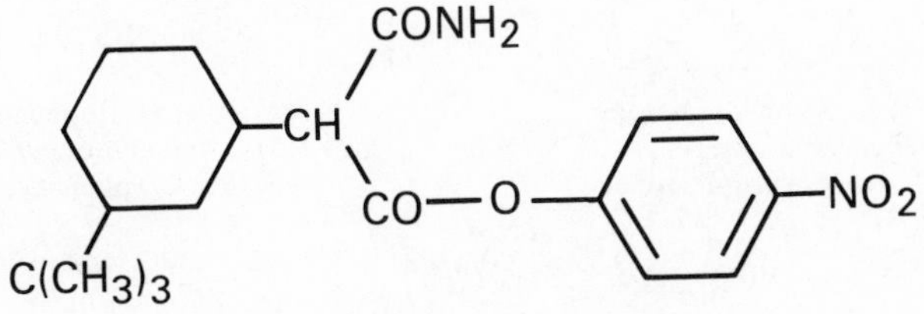

4-nitrophenyl 2-(3-*tert*-butylcyclohexyl)malonamate [IUPAC]
cyclohexaneacetic acid,3-(1,1-dimethylethyl),α-aminocarbonyl,4-nitrophenyl ester [CAS]

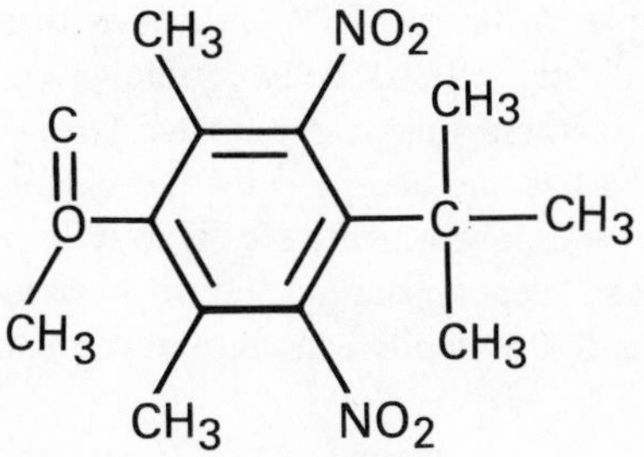

4′-*tert*-butyl 2′,6′-dimethyl-3′,5′-dinitroacetophenone [IUPAC]
ethanone,1-[4-(1,1-dimethylethyl)-2,6-dimethyl-3,5-dinitrophenyl]- [CAS]

Figure 3

of problem. In any case, names such as acetic acid and trinitrotoluene are not going to be abandoned by industry in the near future and they may well be preferred for communication between chemists.

Although the International Union of Pure and Applied Chemistry (IUPAC) has used a philosophy compatible with that of the CAS in their development of rules, the approach has been less rigorous. It codifies established practices but tries to regularise procedures within them. Some of the IUPAC rules are firmer than others and their choice restriction is achieved in part by rejecting archaic, unclear and otherwise deplorable styles and usages. The list of names for the compound in Figure 2 is effectively reduced by IUPAC Rule C-10.41. This

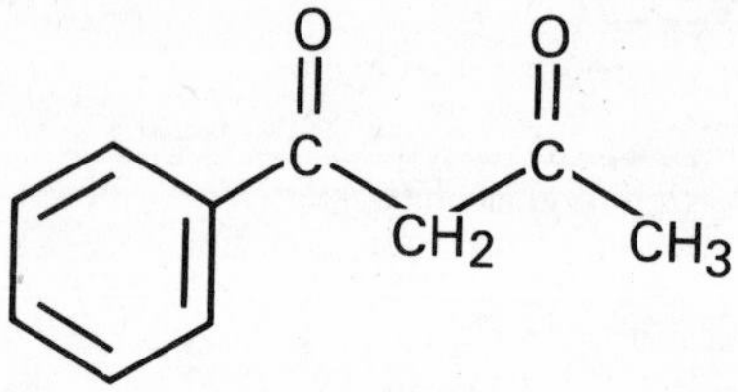

(1) Acetonyl phenyl ketone
(2) Benzoylacetone
(3) 2-Acetylacetophenone
(4) Acetoacetophenone
(5) 1-phenylbutane-1,3-dione
(6) Methyl phenacyl ketone

Figure 4

establishes a firm hierarchy of suffixes and eliminates all these names except the one based on nicotinic acid. However, it is possible for one substance to have several IUPAC names. Butan-1-ol has already been mentioned and Figure 4 gives a further example. Although IUPAC, through their various Nomenclature Commissions, are the only official body promulgating international rules for chemical nomenclature, these rules carry no legal force and they will endure only in so far as they become accepted by the chemical community and its associates in industry and government, etc. However, new chemicals and even new types of chemical keep appearing and so new rules have to be devised to cope with them and this work can have repercussions on names already established.

The advisory service set up by the Laboratory of the Government Chemist (LGC) covers all aspects of chemical nomenclature for use in-house, and for government and associated bodies, such as the British Pharmacopoeia Commission (BPC) and the British Standards Institution (BSI). Some assistance is also provided to industry, commerce and educational bodies. Adherence to the IUPAC rules forms the basis of the advice since this can leave room for choice and certainly for interpretation. How should an advisory service act in such a situation? Various factors can exert influence.

(1) Toleration of existing practices with a view to minimising change. This does not sound very progressive, but it must be remembered that any alteration to

a name, large or small, generates a new synonym and may involve reprinting everything, from an Act, requiring fresh approval by Parliament, down to every notice, label and catalogue entry. Change for its own sake is not merely idle but mischievous. It is not surprising that there is a demand for one name per compound although the basic problems remain in that every user wishes that name to be the one that is used currently in his organisation. The LGC

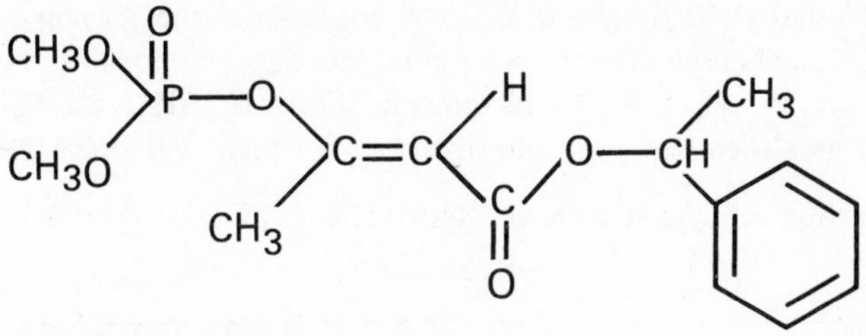

CROTOXYPHOS (ISO)

(a) (*E*)-Dimethyl 1-methyl-2-(1-phenylethoxycarbonyl)vinyl phosphate
(b) 1-Methylbenzyl 3-(dimethoxyphosphinyloxy)isocrotonate

Figure 5

advisory group has found itself at the interface between the academy and the market-place, commanding an overall view. Staff are mindful of the practical consequences of re-naming but at the same time appreciative of the need for system and consistency.

(2) 'Horses for courses'. Should each specialised field be allowed to develop its own nomenclature? This may be possible to some extent. For example, it may not be too serious if BSI and the International Organization for Standardization (ISO) name all organo-phosphorus pesticides as phosphate or phosphorothioate esters. However, it should be appreciated that this would not produce an IUPAC name in every case. Thus CROTOXYPHOS, illustrated in Figure 5, may, by means of name (a), fall into line with other organo-phosphorus pesticides; but this would be a violation of IUPAC Rule C-10.41 which puts neutral esters of inorganic acids below carboxylic acids, thereby requiring name (b).

(3) The IUPAC principles. An advisory service on nomenclature must not disregard IUPAC guidance without an extraordinary, overwhelming reason – such as a genuine failure in the rules. Within this fairly loose constraint there is some room for manoeuvre (a) to minimise change and, when it is unavoidable, to soften its effects and (b) where possible, to suit advice to the customer's special needs. Such a line can be taken only with great care as apparently diverse

fields have a way of converging. If, for example, a nematocide named by an ISO Committee comes under the scrutiny of the Pharmacopoeia Commission as a veterinary medicament and different names have been recommended to each body, then some dispute is inevitable. Internal consistency must therefore underlie advice and accessible records must be kept to ensure consistency. It must be stressed that IUPAC is effectively the only major authority on systematic naming and that those working in IUPAC Nomenclature Commissions do well to recognise that their 'target-area' is not confined to the academic world but ranges over a large and diverse user public, all desiring guidance and possibly under pressure to follow IUPAC recommendations. If IUPAC rules fail to cover a particular area then it is probable that no other body will repair the omission.

Compilation of the European Repertoire of Trade Chemicals has been

Formula	Traditional name	IUPAC name
$H_2S_2O_7$	pyrosulphuric acid	disulphuric acid
$Na_2S_2O_7$	sodium pyrosulphate	sodium disulphate
$NaHSO_4$	sodium bisulphate	sodium hydrogensulphate
$NaHSO_3$	sodium bisulphite	sodium hydrogensulphite
$Na_2S_2O_5$	Sodium metabisulphite	sodium disulphite

Figure 6

part of LGC's work for the last seven years. This list covers Euro-chemicals of trade-interest and has grown continuously since 1914. It developed originally when it was discovered at the outbreak of hostilities that the dyestuffs industry, essential for such items as uniforms, was in enemy hands. The so-called 'Key Industry Duty' list of chemicals was produced to assist the hastily formed fledgling British chemical industry at that time. This list has grown continuously and successive editions of this book[2] giving tariff headings for each chemical, have used all the nomenclature styles of the period 1914–1974. This last was the year when work began to combine the original information with the European tariff list and to reform and unify the nomenclature under the IUPAC rules. In order to reduce the choice which IUPAC often provides, it has been necessary to impose *ad hoc* rules and here decisions have been recorded together with the associated reasoning. Attitudes can be modified as shown in Figure 6 by the example of the condensed sulphur oxo-acids. The difference between 'di' and 'bi' in this example is likely to prove too subtle for non-chemists and possibly even for some chemists. Under practical conditions, confusion would seem almost inevitable, and therefore the traditional names were recommended even though IUPAC have aimed at consistency in the approach to Group VI elements. (It is unlikely that a user would quote 'sodium pyrochromate' for $Na_2Cr_2O_7$ rather than 'sodium dichromate', although the older bichromate-synonym may reinforce some misgivings.) However, the IUPAC names seem to have been accepted without great problems.

Concern for practical problems must be balanced against reforming zeal and reform of nomenclature should be nursed along gently in the direction of system

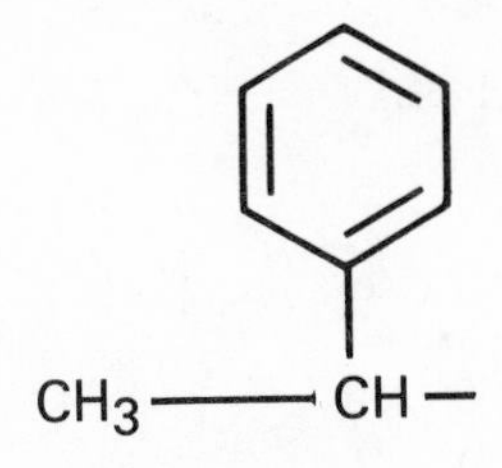

1-phenylethyl- or α-Methylbenzyl- ?
Figure 7

and international acceptability. This is of particular importance given that the style of names in other languages is often determined by that of the English original.

The *ad hoc* sub-rules used by the LGC advisory service have had to be numerous and it is hoped that they are logically based. On occasions the choice has had to be arbitrary. In the example in Figure 7 there are strong grounds for using 1-phenylethyl although pros and cons are finely balanced and the IUPAC rules are indecisive on this point. A further example is given below:

Inorganic IUPAC Rule 6.2: sodium hydrogencarbonate
Organic Rules IUPAC C–462.1: sodium dihydrogen citrate
Adenosine 5′-(disodium dihydrogen triphosphate)

The IUPAC Inorganic Rules (the so-called 'Red Book') join hydrogen to sulphate whilst the Organic Rules (the so-called 'Blue Book') use ethyl hydrogen sulphate in three separate words – a tiny variation which results in problems out of all proportion to its importance. Some users would argue that a name is essentially the same either way but compilers of lists have to know whether there is a space or not. If there is freedom in the matter, a computer will list both as synonyms.

2,2′-dithiodiacetic acid [dithioglycollic acid]
[dithioglycollic acid] **2,2′-dithiodiacetic acid**
[dithiol] **toluene-3,4-dithiol**
dithio-oxamide [rubeanic acid]
[dithiosalicylic acid*] **di-(2-carboxyphenyl) disulphide or 2-hydroxydithiobenzoic acid**
dithizone
dithymol di-iodide
***NN′*-di-*o*-tolyguanidine**
***NN′*-di-*o*-tolylthiourea**
di-(3,5,5-trimethylhexyl) phthalate [dinonyl phthalate]
di-1,2-ureylene-ethane [acetylene diurea]
1,4-divinylbenzene [*p*-vinylstyrene]
divinyl ether
dodecanal (lauraldehyde)
dodecanedioic acid (decane-1,10-dicarboxyl acid)
dodecane-1-thiol [dodecyl mercaptan]
(dodecanoic acid) **lauric acid**
dodecan-1-ol [lauryl aclohol†]
[dodecyl mercaptan] **dodecane-1-thiol**
dodecyl sodium sulphate (sodium lauryl sulphate)
dulcin [*p*-phenetylurea] [*p*-phenetylcarbamide]
[durene] **1,2,4,5-tetramethylbenzene**
[*iso*durene] **1,2,3,5-tetramethylbenzene**

emetine
ephedrine
[epichlorohydrin] **1-choloro-2,3-epoxypropane**
[Epsom salts] **magnesium sulphate**
ergometrine
ergotamine
ergotoxine
[erythrite] **erythritol**
erythritol [erythrite]
erythritol tetranitrate
[eserine] **physostigmine**
ethanediol [ethylene glycol]
ethanethiol [ethyl mercaptan]
ethanol [ethyl alcohol] [alcohol]
ethanolamine (2-hydroxyethylamine)
[ethanolamine phosphoric acid, barium salt] **2-aminoethyl barium phosphate**
ethanolamine *N*-oleate
ethanolamine *O*-oleate

*this name has been used indiscriminately for both compounds.

†The lauryl alcohol referred to here is **not** the commercial mixture of fatty alcohols generally known as 'lauryl alcohol'.

Figure 8 – Extracted from BS 2474:1965. (Reproduced by permission of the British Standards Institution, 2 Park Street, London W1A 2BS from whom complete copies of the standard can be obtained)

Indeed, where a name has n points of alternative choice (e.g. brackets or no brackets), the total possible number of synonyms generated will be 2^n. The decision taken by LGC on this issue has been to close up the name for inorganic substances but not for organic substances. However, this still raises problems on products such as adenosine 5′-(sodium dihydrogen phosphate). Biochemicals are organic in composition and are left with a space. The important point is that whenever a decision is taken, then it should be recorded and followed for future decision making.

There are a number of publications which provide helpful guidance for use by industry and government. Information in BS Specification 2474:1965, *Recommended Names for Chemicals used in Industry*, was completed with guidance from LGC's advisory service to the responsible Committee which had representation from both industry and Government. The Standard is used as a source book when chemical names are cited in legislation, although it does not purport to be as comprehensive as the Euro-repertoire. A typical page is reproduced in Figure 8. The first-preference names are printed in bold type, respectable synonyms are included in parentheses and deprecated names in square brackets. New editions will indicate the general trend for progressive names as older names slip first into the ( ) and then the [ ] in successive editions and are ultimately dropped from the list. Those engaged in the compilation of lists of chemical nomenclature must remain astute observers of the industrial scene and be able to strike a balance between cognisance of academic progress towards systematic reform and recognition of the industrial facts of life. Many industrial firms are progressive in their use of chemical nomenclature and the work of the BSI must not be allowed to lag behind industrial modernity. However, practices too much in advance of their time do not gain acceptance and attempts to force the pace can often result in the entrenchment of conservative opposition and delay eventual progress.

BS 2474 omits pharmaceuticals, dyestuffs and pesticides as these are comprehensively covered in other publications. In the case of pesticides and pharmaceuticals, users are in a fortunate position The Nomenclature Commissions deliberate on the selection in both fields of short, distinctive approved names (the so-called INNs) of the World Health Organization (WHO) and those of ISO. Such names are devised for international use and whenever possible are recommended for official legislative purposes in United Kingdom or EC contexts. Systematic names under the IUPAC Rules are chosen where an INN has not been promulgated for a substance.

To summarise, chemical nomenclature used for official purposes is a mixture of the best available international advice, adapted with a knowledge of current use in industry and designed to achieve the widest understanding of the requirements of legislation.

REFERENCES

1. *Naming and Indexing of Chemical Substances for CHEMICAL ABSTRACTS during the Ninth Collective Period (1972–1976)*, American Chemical Society, Columbus, Ohio, 1973.
2. *Classification of Chemicals in the Brussels Nomenclature and in the UK Customs and Excise Tariff*, HMSO, London.

CHAPTER 7

The role of the United Kingdom in chemical nomenclature

Dr H. EGAN
Laboratory of the Government Chemist, London, United Kingdom

The principal aim of chemical nomenclature is the retrieval and unambiguous communication of information relating to identifiable chemical substances in, so far as it is compatible with the other objectives, as simple a manner as possible. Chemical nomenclature today is an international matter and it is difficult to say that the United Kingdom has any unique role or function in the field. It has of course contributed fully, and continues so to do, to the international efforts in this cause, not least to the efforts of the International Union of Pure and Applied Chemistry (IUPAC) Interdivisional Committee on Nomenclature and Symbols, its associated nomenclature commissions and the joint nomenclature commissions of IUPAC with other International Council of Scientific Unions (ICSU) members. It is arguable that English is the principal basic language of chemical nomenclature, even having due regard to the Latin, Greek and other classical origins of many of the elements (if that is not too confusing a word in this context) of nomenclature. This is not to ignore the extensive development of chemical nomenclature in French, Italian and other European languages, or the corresponding Chinese, Japanese, Russian and other enterprises in other scripts, nor indeed the differences that exist between American English and the Queen's (or King's) English (as it is called in Britain). The majority of published scientific papers are in the English language, as are the publications of the American Chemical Society and its Chemical Abstracts Service. The latter is a particularly important area of influence in chemical nomenclature and is the only indexing and retrieval system in the field with any claim to near-comprehensiveness. Perhaps the United Kingdom members of IUPAC nomenclature commissions unconsciously safeguard the essentially English grammar of nomenclature, notwithstanding the many by-ways still developing which have left conventional grammars far behind. On the other hand, in earlier days those speaking English invented perhaps more than their fair share of trivial, totally unsystematic names of which Dingler's Green, batyl alcohol and Count Palmer's Powder have been quoted as typical examples.

Although many attempts were made to develop systematic chemical nomenclature, at least in selected areas, before 1850, Britain was not initially notably in the forefront of this field. The recognition of the problem and the need for a systematic approach was probably first realised in modern times by Dr H. E. Armstrong, who just over a century ago wrestled with the question of recording the configuration of simple di- and tri- substituted benzene isomers in the new series of the *Journal of the Chemical Society* which had commenced in 1876. This is evident from the extensive footnotes to a 37-page review of recent work on the isomers of aromatic amines and other substituted benzene compounds which he initialled in 1876[1]. The original attributions in Germany of the designations *ortho-*, *meta-*, and *para-* had proved to be incorrect and Armstrong was at pains, in his abstracts, to put the nomenclature of the compounds concerned on a correct and systematic basis. There is a reference in the next volume to a lecture by him on the subject of systematic nomenclature[2] but apart from a passing reference several years later[3] there is unfortunately no other record of this.

Considerations earlier in the century by Berzelius, Dumas and others on the European continent had provided the basic language from which the nomenclature for organic chemistry could develop[4]. The matter had been raised by Daubenoy at a meeting of the British Association in 1851[5], when he expressed the view that many of the names having 'already extended to the limit of ready utterance', they should be confined in length to no more than six or seven syllables. Guidance on nomenclature to contributors and abstractors was given by the council of the Chemical Society in 1879[6]. The Geneva Congress of 1892[7] represented the first major step towards international agreement on chemical nomenclature, when Professor Armstrong, now joined by (Sir) William Ramsay and Dr J. H. Gladstone from Britain, contributed substantially to various areas of the debate. Some aspects were, owing to shortage of time, treated less thoroughly than they deserved to be but the way was then open, shortly after the formation of the International Union of Chemistry, for Sir William Pope (President of the Chemical Society 1917–1919) in 1922 to propose the formation of a committee composed of delegates from the staffs of the leading chemical journals to look further into the problems. So began the Interdivision Committee on Nomenclature and Symbols of IUPAC; and the comprehensive, though still developing, IUPAC system of nomenclature for chemical compounds which is used today.

After a considerable period of international interaction and consolidation, the editor of the *Journal of the Chemical Society* lectured to the Society in 1936 on the subject[8] but it was left to his successor Dr A. D. Mitchell to present what he entitled *British Chemical Nomenclature* in a more formal monograph twelve years later[9]. At much the same time Dr Malcolm Dyson was pioneering his linear notation system[10].

There is one other important area of United Kingdom influence in the field of chemical nomenclature, that of the *Journal of the Chemical Society* itself and its successive editors. This journal was first published in 1848 and I have already mentioned the guidance given by council in 1879. In addition to Mitchell (Assistant Editor 1926, then Editor) it is proper here to recall the work of R. S. Cahn and Lionel Cross in the 1950s and 60s.

There are several organisations active today in the development and use of chemical nomenclature in the United Kingdom. The Royal Society of Chemistry (originally the Chemical Society) has been a major pioneer in the field of systematic chemical nomenclature, in recent years through the United Kingdom Chemical Information Service (UKCIS). The British Pharmacopoeia Commission's Nomenclature Committee maintains and publishes a list of British Approved Names (BAN) which is the ultimate basis for the identity of a wide variety of substances used in medicine. BS 2474: 1965 *Recommended Names for Chemicals used in Industry,* published by The British Standards Institution (BSI), is the work of BSI Technical Committee CIC/3 which is continuing the task of modernising such names. In this the committee seeks to eliminate archaic and misleading names and to collect useful synonyms, expressing preference for those terms which reflect progress towards internationally agreed systems of chemical nomenclature which have been developed under the auspices of IUPAC. The Standard, which continues to serve as a source-book when chemical names are cited in legislation, is currently being revised by the committee. It is expected that the revised Standard will become the basis for the consideration of an international standard by the International Organization for Standardization (ISO). BSI pioneered this field in the area of common names for pesticides (the latest edition of which also provides advice on the pronunciation of names[11]), which is also being further developed by ISO[12]. In addition to the expertise in many individual universities and institutes, technical and otherwise, many commercial and industrial organisations maintain an expertise in and use systematic chemical nomenclature. Indeed, they contribute to the various national enterprises in this field which have been referred to above.

Central government interest is focussed by the Chemical Nomenclature Advisory Service (CNAS) of the Laboratory of the Government Chemist (LGC)[13]. An early interest in this field was taken by a previous Government Chemist, Dr G. M. Bennett, who collaborated with C. A. Mitchell in writing *British Chemical Nomenclature*[9]. This is also a major route for United Kingdom input into the European Community and other applied areas of international interest, notably the compilation and updating of the Customs publication on the classification of chemicals extending to some 18,000 entries[14]. This listing has been combined with a list of European chemicals to produce a combined 'repertoire' of some 20,000 Euro-chemicals in the six official languages of the European Community[15]. Through LGC the United Kingdom has for many years provided the Chairman of the Chemists' Committee of the Nomenclature

Committee of the Customs Co-operation Council, the organisation which is responsible for the Customs Co-operation Council (CCC) Nomenclature (formerly known as the Brussels Nomenclature), the basic system of tariff classification on which most of the national tariffs of the world are based, including those of the United Kingdom and the European Community. The 99 chapters of the CCC Nomenclature cover all goods in international trade, including in detail the entire chemical industry. The chapters for organic and inorganic chemicals are subclassified essentially according to chemical composition but with due regard for chemical nomenclature. It was not until 1979 that the attitude of the British National Committee for Chemistry towards nomenclature, particularly in relation to the international aspects of the subject, was formalised with the formation of the joint Royal Society–Royal Society of Chemistry Panel on Chemical Nomenclature. The terms of reference of the Panel are

(1) to consider (or be responsible for expert views to be obtained in respect of) proposals for chemical nomenclature and associated terminology and conventions, including their implications in general and in the United Kingdom in particular;

(2) to consider the application of the IUPAC recommendation in the United Kingdom, including problems arising from different interpretations of such recommendations.

Its membership embraces representation of all the United Kingdom professional societies with a direct interest in chemistry, of industrial and educational interests and of United Kingdom membership of the various IUPAC nomenclature commissions including Interdivisional Committee on Nomenclature and Symbols (IDCNS). The Chairman and United Kingdom members representing the IUPAC Commissions and IDCNS are appointed by the British National Committee for Chemistry after consultation with the President of the Royal Society of Chemistry and serve for a term of three years in the first instance, with reappointment for a second term of three years if required. Further reappointment is not allowed except after a lapse of two years and appointments lapse if membership of the relevant IUPAC body ceases. If a Nomenclature Commission of IUPAC has no person from the United Kingdom amongst its membership, the British National Committee can take steps to appoint a National Representative who then is eligible to serve as a member of the Panel; pending such an appointment, the National Committee may make a one-year appointment to the Panel of a recently retired past member of the Commission. Nominees of the specified organisations are also appointed for a term of three years in the first instance, with reappointment for a second term of three years, if required. Further reappointment is not allowed except after a lapse of two years.

The joint Secretaries of the Panel are the Executive Secretary of the Royal Society and the General Secretary of the Royal Society of Chemistry. The Panel reports to the British National Committee for Chemistry, through which it communicates with IUPAC on relevant matters; and to the Executive Committee of the Royal Society of Chemistry for the information of the British chemical community. It also formulates, for consideration by the British National Committee for Chemistry and the Executive Committee of the Royal Society of Chemistry, the procedure to be followed in handling any recommendations issued by IUPAC.

Examples of the manner in which the Panel works are the evaluation of tentative or provisional IUPAC nomenclature recommendations in consultation with those in the United Kingdom most competent to comment on the particular areas concerned. It ensures that definitive recommendations from IUPAC or other bodies are brought to the attention of organisations concerned with industry, research, education and legislation in Britain and to encourage their adoption. It also considers United Kingdom representation on international committees concerned with nomenclature matters and is able to advise on nomenclature and where appropriate to pass matters on for international consideration.

REFERENCES

1. *J. Chem. Soc.*, 1876, **1**, 204.
2. Anon., *J. Chem. Soc.*, 1876, **2**, 685.
3. H. Egan and E.W. Godly, *Chem. Brit.*, 1980, **16**, 16.
4. M. P. Crosland, *Historical Studies in the Language of Chemistry,* p. 338, Constable, London, 1962.
5. Report of the British Association for the Advancement of Science 1851, pp. 124–132.
6. Anon., *J. Chem. Soc.*, 1879T, **35**, 276.
7. H. E. Armstrong, *Nature*, 1892, **46**, 56.
8. C. Smith, *J. Chem. Soc.*, 1936, 1067.
9. C. A. Mitchell, *British Chemical Nomenclature,* Arnold, London, 1948.
10. G. Malcolm Dyson, *A New Notation and Enumeration System for Organic Compounds*, Longmans, Green, London, 1947.
11. BS Specification 1831:1969, *Recommended Common Names for Pesticides*, British Standards Institution, London, 1969.
12. ISO Standard 257:1976, *Principles for the Selection of Common Names (Pesticides*), International Organization for Standardization, Geneva, 1976.
13. Report of the Government Chemist 1976, p. 125, HMSO, London, 1977.
14. *Classification of Chemicals in the Brussels Nomenclature and the United Kingdom Customs and Excise Tariff,* HMSO, London, 1973.
15. *Classification of Chemicals in the Customs Tariff of the European Communities,* Commision of the European Communities, Luxembourg, 1981.

CHAPTER 8

The use of chemical nomenclature in Chemical Abstracts

M. G. ROBIETTE
The Royal Society of Chemistry, Nottingham, United Kingdom

This paper is concerned with the criteria that affect Chemical Abstracts (CA) nomenclature, it describes a little of its history and development, comments on the processing of nomenclature within Chemical Abstracts Service (CAS) and discusses the aids available to the users of nomenclature.

There are many different types of nomenclature, each having advantages or disadvantages depending on its use. CA nomenclature is designed for a specific purpose, namely for use in a large printed index. The CA substance index is the largest printed index to chemical substances in existence. It is also the largest index to any printed work and recognised as such by *The Guinness Book of Records*. CAS publishes indexes to each volume and these are periodically collated into collective indexes, originally for ten years but now covering a five-year period. The *9th Collective Index,* usually abbreviated to 9CI, is the last to have been published and contains 7,351,174 substance entries for 1,924,001 different substances. The distribution of entries is very uneven; 1.4 million substances have only one entry. An index to the 10,000 most common compounds, which has sometimes been suggested as a useful additional product, would not be a great deal smaller than the complete index.

A large printed index imposes particular constraints on the nomenclature that can be used and some of the criteria are:

(a) Unique and unambiguous. Probably the most important criterion. Names must be unambiguous, i.e. each name should lead to only one compound, and be unique, i.e. each compound has only one possible name. Many nomenclature systems give unambiguous names but very few give unique names. The systems recommended by the International Union of Pure and Applied Chemistry (IUPAC) often allow alternative names for compounds. This is perfectly acceptable for many purposes but not for indexing; the user must be sure that all occurrences of the same compound are indexed in the same place or the index will be useless.

(b) Economic and consistent generation. CAS has to name over 300,000 new compounds each year. It is important for a cost-effective information service that this should be done as quickly as possible. Different indexers must be able to name compounds consistently without extra time being required for checking. The system should be as simple as possible, without exceptions and difficulties of interpretation.

(c) Easy translation from name to structure. Commonly, users searching for a specific compound start with the molecular formula index. The index may list several compounds with the particular formula. The user then has at least partially to translate these names into structures in order to ascertain whether the right compound has been identified.

(d) Derivable by users. The formula index is only of value when the user is looking for specific compounds. If the user is looking for derivatives of a specific parent structure then he has to be able to derive its name as a prelude to using the substance index. Therefore names must be derivable by users as well as by indexers.

(e) Bringing similar compounds together. There are two reasons for this requirement. Firstly it permits a certain level of generic searching, i.e. looking for compounds of related structure. This is of rather limited value in a printed index. Secondly, the ease of using a list of related compounds is improved if they are located physically near to each other in the index. This reduces the amount of page turning and lessens the movement of books from the shelves. There are some types of nomenclature which are particularly used in indexes because of their ability to bring related compounds together. Conjunctive nomenclature is a type of nomenclature that can be used for compounds containing a ring system and a functional group linked together by a saturated aliphatic chain, e.g. 1-naphthalenebutanoic acid. This type of name brings together compounds with the same ring system; the alternative name butanoic acid, 1-naphthyl- would lose the ring information for sorting purposes. Another example is the method of naming functional derivatives, e.g. esters. By citing the ester information after the name of the acid, e.g. butanoic acid methyl ester, all esters of the same acid are brought together.

(f) Machine searchability. Nowadays many names are searched by computer as well as by humans and this must be borne in mind in the design of the nomenclature system. However, much the same criteria apply to human and machine searching. Machine searching of names is likely to decrease in importance as structure searching systems, e.g. CAS Online[1], become increasingly available.

(g) Consistency with international agreements. It is important when producing names on the scale of CA that they fit with those approved internationally. CAS works very closely with relevant bodies, especially IUPAC. CA names conform with IUPAC principles but with additional features, e.g. for uniqueness. There is a problem in that names for new types of compounds sometimes have to be assigned ahead of international discussion, but informal liaison normally copes with this difficulty.

Nomenclature policy has to remain constant within a collective period. In the first collective period (1907–1916) compounds were indexed using the names given by the authors. This led to situations where the same compound appeared at more than one index name, e.g. acetonitrile, methoxy- and glycolonitrile, methyl-. From then on, there was a continuous move towards greater systematisation and rules leading to unique names. At the end of each collective period, rules were reviewed to take account of any problems and changing practices.

The largest change in recent years has been that between the 8CI, which ended in 1971, and the 9CI, covering the period 1972–76[2]. By the end of 8CI, the nomenclature system had grown to be one of considerable complexity. The rules covered 3,100 typed pages and the amount of staff time involved in nomenclature was increasing. For the 9CI, therefore, the system was simplified, with the rules decreasing to 1,800 pages. One simplification was to reduce the number of trivial names used as the basis for names. In the 8CI, a large number of trivial names had been used and this led to problems in selecting the preferred name. This can be illustrated by the compound in Figure 1 which in the 8CI could have been named as a derivative of lactic acid (containing the hydroxy group), butyric acid (a larger acid), or hydrocinnamic acid (containing the ring system). Establishing the preferred name required considerable effort. In the 9CI, the rules very quickly give the parent name as benzenebutanoic acid (the largest parent containing both the ring and the principal group).

The series of compounds in Table 1 gives a good example of the difference between the 8CI and 9CI names. These are four of the possible isomers of

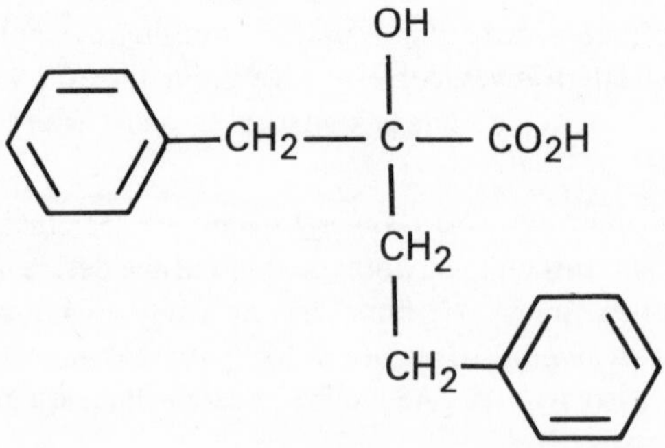

Figure 1

dihydroxybenzoic acid. The 8CI names, on the left, are based on trivial names; the 9CI names, on the right, are systematically named as derivatives of benzoic acid. This illustrates two of the principles of index nomenclature:

(i) the need to bring similar names together: names in the 9CI are located in the same part of the index instead of four different places as used in the earlier edition; and

(ii) the ease in achieving a translation from the name to a structure; all that is required in the 9CI is a knowledge of benzoic acid and how it is numbered.

Table 1

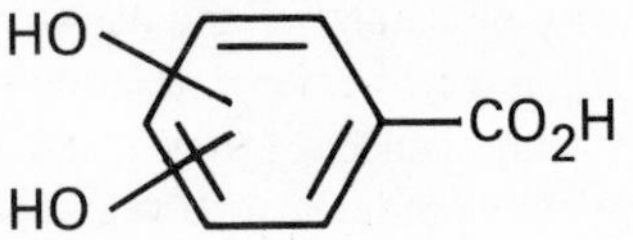

8CI	9CI
o-Pyrocatechuic acid	Benzoic acid, 2,3-dihydroxy-
β-Resorcylic acid	Benzoic acid, 2,4-dihydroxy-
Gentisic acid	Benzoic acid, 2,5-dihydroxy-
Protocatechuic acid	Benzoic acid, 3,4-dihydroxy-

In 8CI a user would have to remember, or look up, the structure of each compound.

The increased use of systematic nomenclature has some disadvantages: names tend to be longer than those for corresponding trivial names. Another problem is that relatively unfamiliar (at least at first sight) names are used for familiar compounds. Some 9CI names are benzenamine (aniline), benzenemethanol (benzyl alcohol), and benzene, ethenyl- (styrene). Although unfamiliar, these names are not particularly difficult, and are not new in concept. They use nomenclature principles that have existed for many years and are fully in accord with international practice.

CAS has now entered the 10th collective period. There have been no significant changes between 9CI and 10CI. Minor changes are planned for 11CI which will start in 1982. Current policy takes account of the large archival files of names which cover a wide time period. Continuity of nomenclature is becoming more important than the need to make short-term improvements.

A key element in the processing of nomenclature is the CAS Chemical Registry System. This consists essentially of two large computer files, one containing names and the other structures. The names and structures for the same compound are tied together with a unique number, the CAS Registry Number. At present there are more than five million structures and eight million names on file. The name file contains both trivial and trade names as well as index names and is therefore larger than the structure file.

The use of this system in indexing is illustrated by Figure 2. When a compound is selected for indexing, an indexer will first input a name for that compound to see whether it matches one already on file. Typically this input name would be a trivial name rather than a fully systematic name. If there is a match, the system will retrieve the Registry Number and the index name which can then be reviewed to see if the match is correct. If there is no match, a machine-readable version of the structure of the compound is generated and matched against the structure file. If there is a match, the number and name are retrieved as described earlier. If the structure does not match then the compound is new to the file. It is assigned a new number and has an index name generated by a nomenclature specialist. Thus a name for a new compound is generated on one occasion only. The index name will be retrieved if it has appeared before. Out of about 1.8 million compounds indexed in a year, about 330,000 are new, so the saving is substantial.

Index names can be checked by passing them through a series of nomenclature translation programmes which generate a structure (connection table) from

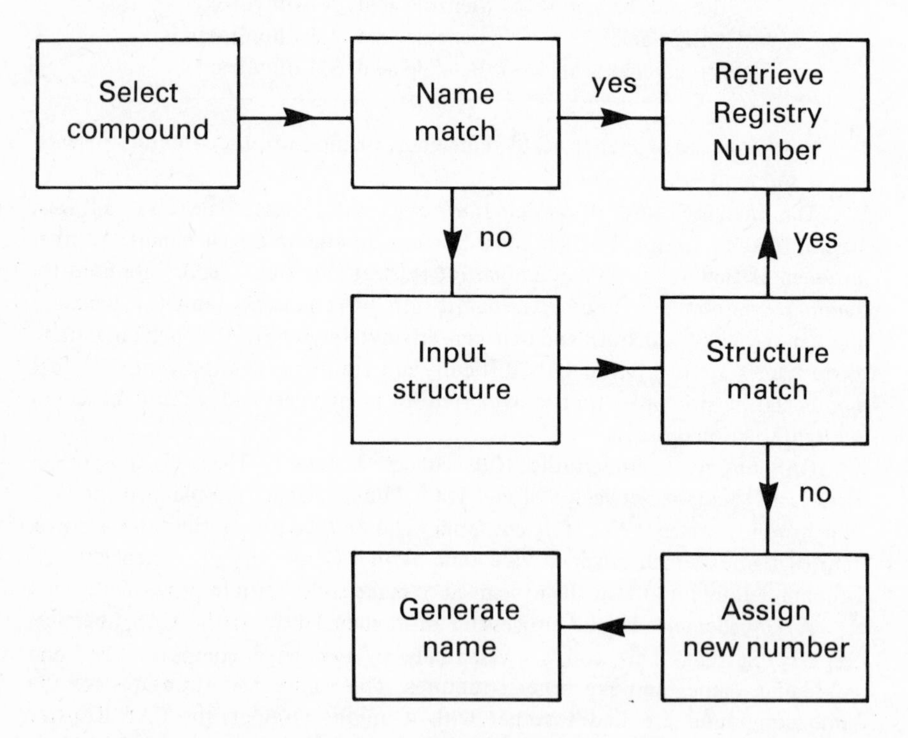

Figure 2 – Substance processing.

the name (Figure 3). This structure can then be compared with a structure already on file and this helps ensure consistent naming.

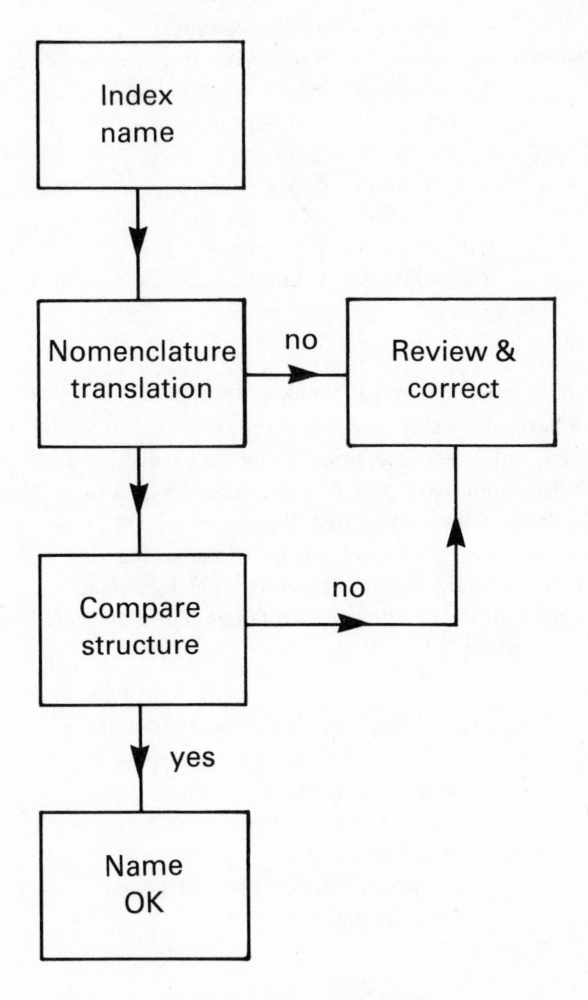

Figure 3 – Name verification.

The Registry Numbers used in this system do not in themselves have any meaning: they are assigned chronologically as new compounds are added to the file. Similar compounds will not in general have similar numbers. Numbers are very specific in that different stereochemical forms of a compound, or labelled compounds, will have different numbers. In Table 2 the numbers on the left are the Registry Numbers, recognisable by the two hyphens and the 2-digit-1-digit pattern at the end. (The last digit is a check digit which identifies transcription errors.)

Table 2 – Registry Numbers

25167-67-3	Butene (unspecified)
106-98-9	1-Butene
107-01-7	2-Butene
624-64-6	(*E*)-2-Butene
590-18-1	(*Z*)-2-Butene
18932-22-4	1-Butene-3,3-d_2
68761-07-9	1-Butene-1,2-$^{14}C_2$

Because they offer a concise, unique description of a compound, CAS Registry Numbers are being used increasingly as compound identifiers in government files, online services, printed reference works, and company internal files. Table 3 lists some examples of publications which make use of Registry Numbers. The *Index Guide* contains information needed to use the indexes and a large number of cross-references from trivial names together with the Registry Number. The *Index Guide* has an Appendix[3] which contains a detailed description of CA nomenclature including various tables and references. This is available as a separate publication.

Table 3 – Publications Using Registry Numbers

Aldrich catalogue
British Pharmacopoeia
Merck Index
Journal of Organic Chemistry
Pesticide Manual

Another publication of particular help to the searcher is the *Parent Compound Handbook*[4] which is useful mainly for finding names of ring systems. This is particularly important as ring systems are often used as a basis for the index names. For each ring system is given the name and diagram, Registry Number, formula and data on component rings and Wiswesser Line Notation. The *Handbook* contains all these features and includes a permuted index to rings that are components of other ring systems.

A number of nomenclature aids are also available which use the Registry Number as an identifier. *The Registry Handbook,* Number Section will provide the index name and molecular formula from the Registry Number. Compounds with trivial and trade names can be found from the *Registry Handbook,* Common Names Section (available in microform). This lists the Registry Numbers and molecular formulae corresponding to trivial names. This enables the searcher to move from the number section for the same product to a list showing the index name and all the trivial names that are held on the file. This facility is particularly useful in that it links together multiple names for the same compound.

CA nomenclature is designed for a specific purpose, namely the production of printed indexes. The constraints brought about from this requirement are likely to remain the same for as long as printed indexes last. There has been speculation that printed indexes will disappear and be replaced by computer-based systems. Although computer systems are becoming more common there are many users who do not have access to such facilities nor have the resources to use them. These include universities, small companies, and developing countries. For this reason, printed indexes are likely to last for some time to come. The growth of large files of names on computer-based systems is likely to inhibit changes in nomenclature. Major changes in index nomenclature are only likely to occur if a new, unified, and completely systematic nomenclature was devised and accepted by the chemical community. A number of researchers have investigated this possibility but it is a matter for speculation as to whether this is likely to happen while printed indexes are available.[5]

REFERENCES

1. N. A. Farmer and M. P. O'Hara, *Database,* 1980, **3**, 10.
2. N. Donaldson, W. H. Powell, R. J. Rowlett, Jr., R.W. White and K.V. Yorka, *J. Chem. Doc.*, 1974, **14**, 3.
3. *Selection of Index Names for Chemical Substances,* Appendix IV, Chemical Abstracts 1977 Index Guide; Chemical Abstracts Service, Columbus, Ohio.
4. J. E. Blake, S. M. Brown, T. Ebe, A. L. Goodson, J. H. Skevington and C. E. Watson, *J. Chem. Inf. Comput. Sci.*, 1980, **20**, 162.
5. A. L. Goodson, *J. Chem. Inf. Comput. Sci.*, 1980, **20**, 167.

CHAPTER 9

The problems of using chemical nomenclature to develop an information base

Dr W. G. TOWN
European Community Joint Research Centre, Ispra, Italy

The desirability and the necessity of handling structural identities with various degrees of sharpness or fuzziness is a major consideration when developing a new data bank. This difficulty is one of the range of problems that had to be overcome when the European Community (EC) commissioned its Joint Research Centre at Ispra to develop a new information bank on chemical substances for use by member countries.

The Environmental Chemicals Data and Information Network (ECDIN) programme was started in the 1970s as a project of the Commission performed both in the Joint Research Centre and also by contract research. ECDIN is conceived as a databank for users concerned with environmental chemicals and the management of these substances. The need for such a system developed as a result of the legislation concerning chemical substances.

The main users of ECDIN will be administrators, at all levels, both in the Commission and in Member State Governments and local governments, and in particular their scientific advisers. There are many other potential users for such a system, including research and development organisations, universities, individuals, chemical companies, etc.

The design for ECDIN in the ADABAS Data Base Management System is shown in Figure 1. ADABAS is a piece of commercially available software which allows the Centre to establish a number of interconnected and related files of information. Each of the rectangles shown in the figure represents a file in the ECDIN ADABAS data base. Two-thirds of the files had been developed by 1981. The relevant files relating to chemical nomenclature are the compound file which is central to the whole data base; the names file which contains chemical synonyms for these compounds, and the chemical structure file.

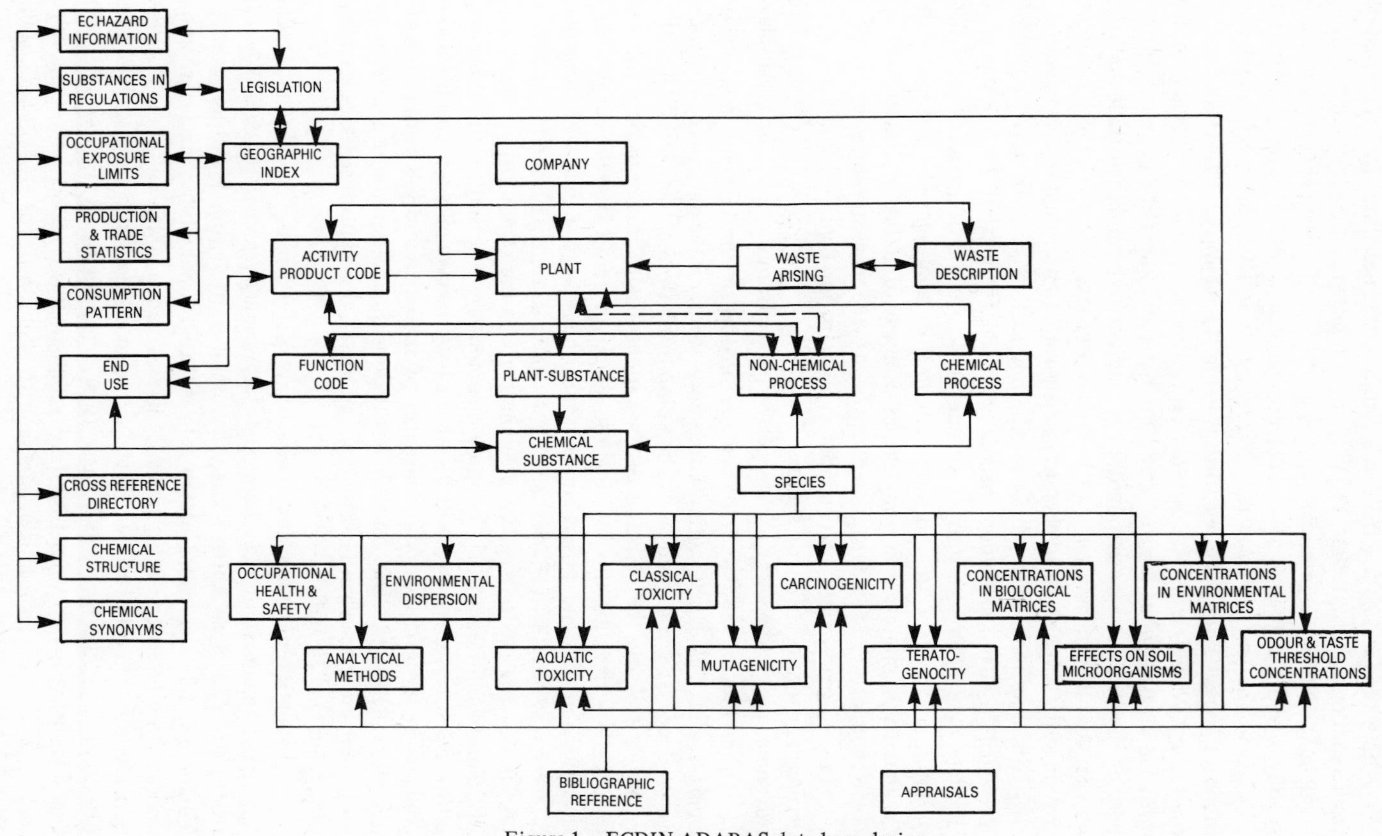

Figure 1 – ECDIN ADABAS data base design.

A data bank can be described as a computerised handbook to differentiate it from a bibliographic data base. ECDIN can be considered as a set of computerised handbooks which are interrelated. The part of ECDIN which consists of the synonym and compound files could be regarded as a computerised chemical dictionary. The part which relates the compound to its toxicity and the relevant bibliographic data could be regarded as being similar to the NIOSH *Registry of the Toxic Effects of Chemical Substances*. However, the toxicity data is recorded with far greater detail than is found in that particular publication. A further part of the data base corresponds to a handbook of published trade statistics on chemical substances. A user of ECDIN will have access to a set of handbooks covering different aspects of chemical substances in one complex data base.

Two questions arise when considering ECDIN. Firstly why is the EC interested in chemical names? Secondly should chemical structures appear in the data base? The files provide an important method of accessing other files of information; they are also interesting in their own right but principally the Community is interested in using names as a way of accessing data on environment chemicals. The same argument applies to chemical structure information.

The current names file includes synonyms for some 25,000 substances originating from a variety of sources. The source of each synonym is tagged to the name so that the user can pull specific names from the data base for display.

Other files in the data base contain information on chemical names. For example, a file called 'EEC hazard information' contains the dangerous substances list produced by the Commission. The Commission through the work on ECDIN, is moving towards the collation of the various lists of chemicals on legislation to help those needing this information. Work on the ECDIN Project does not have the aim of standardising nomenclature because that does not form part of the terms of reference. The target is to provide users with the possibility of juxtaposing the names used in different parts of the Commission's work. The EEC hazard information file is simply a computerised version of the dangerous substances list produce by a different branch of the Commission. This list is also available in the seven working languages of the Community. Greek names are not yet available in ECDIN for technical reasons. In addition to the general name files, there are special status name files and these can be pulled out as special sets of the data on demand.

Another file has been devised as a 'cross-reference directory'. The SANSS data base, that part of the Chemical Information System of NIH/EPA which relates the Chemical Abstracts registry number to various identifiers used in handbooks, data collections, etc., is held on ECDIN. Sometimes this information indicates the presence of the substance in a particular collection; sometimes the data contains a local identifier. These act as access points to the data base and, just as in CIS, the user will be able to search with the RTECS number, for example, or with a code used by the Crystallographic Data Centre in Cambridge and find the substance in the data base. In addition to the SANSS data, some

European collections have been entered which are of specific interest to Ispra staff and to future users of the data base.

Each file is a potential access point in the data base. Searches on almost every field in every file in this data base can be made using the standard software which comes with the ADABAS. Questions can be raised for example on substances having a particular melting point; a particular LD50 or range of LD50s; their listing in a particular directive; or a specific end use. The searches can be combined in a completely flexible way. Unfortunately the use of the standard software requires a certain expertise in using the software itself and secondly a detailed knowledge of the data base. The ECDIN Display System is being developed in parallel with the standard package and it is this system which will be used for the first ECDIN experimental service on Euronet DIANE.

Search facilities using the ECDIN Display System will be limited initially to compound-based searches. In the first version there were three entry points: the name, the Chemical Abstracts Service (CAS) Registry Number and the ECDIN number. The concept of the system is simple but it will answer most of the questions expected to be put to the data base. User surveys undertaken in the late 1970s found that approximately 80-90% of the questions formulated by potential users were compound-based. Development effort has concentrated on this type of access for the first phase.

Once the substance has been found then the facilities provided by the display system can be used to select the data which is of interest to the user. The system is designed to be user-friendly and the later version will allow more complex navigation through the data base.

The concern for chemical names is quite different from that of most nomenclature specialists. Nomenclature is used as an access method to data. There is no interest in deriving a unique preferred systematic name although a name of this type is present in the data base and this is the latest preferred index name of CAS. There are many synonyms in the data base. It is expected that all the synonyms available in the CAS Registry System for 70,000 substances will be held on file. The use of these synonyms still has to be negotiated with CAS and will be subject to an agreement with that organisation. The synonyms have been made available to the EC in the context of the work on the EINECS inventory, and, subject to agreement, will be made available to users of the ECDIN data base.

The name used when searching the file does not have to be the preferred name. A phonetic name search prodecure has even been developed which operates whenever an entered name does not appear as one of the synonyms in the data base. ECDIN is one of the first systems to introduce a search of this type, although CIS have since implemented something similar. An example of this type of search is a user accessing the word 'feenol' when 'phenol' should have been used. In this instance the system will reply that it has been unable to locate the substance 'feenol' spelt in this way but it has found five substances with similar sounding names. Thus the data base of some 25,000 substances has been reduced to five and each of these can be examined to decide whether the substance is of interest to the user. In this example one of the first substances retrieved is phenol and an interesting point is that this mistyped name has the phonetic equivalent of three of the synonyms in the data base. Interestingly the Italian name comes out with the same phonetic characteristics as the English name. Some noise has been experienced when, for example, the substance 'vanilla' is found during search. This noise occurs because the phonetic search is conducted by completely disregarding vowels – it works only on the consonants. It makes equivalents between certain consonants; 'f' and 'v', for example, have the same phonetic value. The 'l', because it occurs twice consecutively, is only counted once for its phonetic value and so on. This can be a very useful tool, particularly in the area of trivial names/generic names. This particular phonetic algorithm is an established feature of ADABAS but has been adapted in a new way for use with the ECDIN data system. Future plans include further work on this algorithm to make it more useful for systematic chemical nomenclature.

Other software relevant to chemical nomenclature is held at Ispra. CROSSBOW has been available since 1975 and used for checking the Wiswesser Line Notations in the data base, fragment generation, and diagram generation, and for substructure search. Work is being undertaken on integrating the CROSSBOW system into the ADABAS data base. At present CROSSBOW is a stand-alone system but in the near future it will be operated from files which are themselves part of the ADABAS data base and there will be the possibility of communication between the CROSSBOW system and the ECDIN display system in order to combine searches. This will not be available in the first release of the ECDIN experimental service. The DARC system is also available at Ispra. This is a more sophisticated system, but as it is a 'black box' it cannot be opened and any effort has to be concentrated on developing the use of CROSSBOW.

Another system partly funded by the ECDIN Project, is the work done by Ugi in Munich. It is a system originally designed for reaction prediction covering such features as the design of synthetic pathways. Ugi has been trying to develop the use for predicting the degradation of substances in the environment. These systems could be used to look at structure–activity relationships as part of the future research activities in the context of the ECDIN Project.

The problem of differences in structure representation in relation to substances such as salts and stereochemistry has been considered during the development of the ECDIN data base. This problem can be partly overcome with substructure search but this is not the most convenient way of handling it for the user. Other types of problems exist when handling information which relates to a group of compounds or to a mixture. A modification of the present system under current consideration will allow data to be recorded at different levels. For example, some information about xylenes as a group may be available, e.g. a production statistic or a trade statistic. This information may not be specified down to the individual substance level. Again some information may be held which relates to a mixture, for example a toxicological study on a mixture of xylenes, perhaps specified, perhaps unspecified. There also could be information on the individual isomers. The concept used for ECDIN is that entries are needed for each of these types of situation and for the possibility of relating one set of information to another automatically in the display system. This will enable the user to start a search with 'xylene' and then be told by the system 'this record has been found but you should know that there are three more or four more records relating to xylene, are you interested in including these in your search or not?'. This is the only way that the dispersion of information over different structure identities can be handled. These different degrees of sharpness and fuzziness of structure identity exist in the real world and it is necessary to ensure that they can be handled in any useful data base scheme.

A few examples can be quoted to illustrate this problem: cadmium and its salts, a particular entry in the black list of one of the directives of the Community; PCB, a common group of substances; soluble lead salts which have particular toxic properties not possessed by insoluble lead salts; etc. Information must be recalled at group levels in a data base such as ECDIN and provide links between a group and its members. The present system that has been developed should allow this to be done partly automatically and partly manually.

A few other areas exist where there are similar problems and these include technical grades of substances and groups. These are not well handled by the CAS registry system. Part of the data base in ECDIN should have in its concept a one-to-one relationship to the CA Registry Number and hence to the CA registry system definition of that substance. In addition it is necessary to have entries which relate to groups of substances.

CHAPTER 10

The development of a multilingual chemical repertoire

H. J. CHUMAS
Commission of the European Communities, Brussels, Belgium

A considerable amount of good work has been undertaken already in the field of standardisation of chemical nomenclature. The usefulness of this work could be further enhanced if more account was taken of both commercial and international needs. The European Communities (EC) multilingual chemical repertoire has been developed with this aim in mind. The project can be considered from a number of angles. These include

(a) the need for a multilingual chemical repertoire;
(b) the objective of the repertoire;
(c) the problems encountered in developing the repertoire;
(d) the solutions adopted;
(e) a brief outline of the content and construction of the repertoire;
(f) future developments;
(g) the need for a change in attitude amongst scientists in general and the International Union of Pure and Applied Chemistry (IUPAC) in particular.

(a) THE NEED FOR A MULTILINGUAL REPERTOIRE

The EC is the largest trading power in the world. Member States live and thrive on trade both within the Community and outside. The value of imports and of exports is equal to roughly 25 per cent of the Gross Domestic Product, taking the Community as a whole; and in certain Member States the figure is as high as 50 to 60 per cent. Within the overall trade figures, the chemical sector is extremely important.

From the figures in Table 1 it will be appreciated that there are a lot of chemical products moving into and out of the Community. These chemicals have to be charged with duty, if they are dutiable, or their eligibility for duty relief has to be established if they come into the duty free or reduced duty category. Furthermore detailed trade statistics have to be collected, not only for governmental purposes but also for the benefit of producers and traders who like to keep a close watch on external trade flows. The European Chemical Federation (CEFIC) is also interested in the statistics in relation to cases of possible dumping.

Table 1 – 1979 Community industrial trade

Total import	$286,300m	=	240,600m ECU
Import of chemical products (18.7% of total)	$53,600m	=	45,042m ECU
Import of separate chemically defined elements and compounds (17.8% of chemical products)	$9,520m	=	8,000m ECU
Total export	$386.067m	=	324,426m ECU
Export of chemical products	$74,184m	=	62,339m ECU
Export of separate chemically defined elements and compounds (19.7% of chemical products)	$4,630m	=	12,294m ECU

To meet the needs for chemical products there are some 550 *Common Customs Tariff* lines, 300 tariff suspensions lines, 60 EFTA and GSP reduced lines, and 1,000 statistical (NIMEXE) lines. Each of these tariff and statistical descriptions is published and applied in the seven different Community languages. There are also more than 15,000 chemical products which are commonly traded internationally and these have to be classified for customs and statistical purposes within the framework of tariff and statistical descriptions and codes just mentioned. Declarations for customs are prepared by clerks in the offices of forwarding agents and then checked by customs officials. Individuals undertaking this work often do not have the benefit of expertise in chemistry and do not necessarily understand English. It can be appreciated that clearance of chemicals through customs may create problems. To try to overcome this difficulty, the Commission decided that a multilingual chemical repertoire should be developed.

(b) THE OBJECTIVE SET FOR THE REPERTOIRE

The objective set for the repertoire was quite specific and simple. Namely, to facilitate the task of importers and exporters of chemicals and of the customs staff who have to process their declarations. To meet this objective a system was needed which would enable:

(a) the unequivocable identification of each chemical, commonly traded internationally, for customs and legal purposes;
(b) the establishment of a name, as brief and as simple as possible, for each chemical;
(c) the use of names which would be as similar as possible in the various languages.

(c) THE PROBLEMS ENCOUNTERED IN DEVELOPING THE REPERTOIRE

The repertoire as produced, has some 20,000 denominations of chemical products in six Community languages and also a listing in Spanish. A considerable number of problems had to be overcome during the development of a repertoire of this size.

The starting bases for the project were (a) a limited Community repertoire covering about 8,000 product denominations in four languages, not including English, with their tariff classification in the *Common Customs Tariff*; and (b) a classification of chemicals in the *Brussels Nomenclature* and *HM Customs & Excise Tariff and Overseas Trade Classification in the United Kingdom* in conjunction with Laboratory of the Government Chemist (LGC). This contained the names of the majority of commonly traded chemicals together with their customs tariff classification but was only available in English.

From these base sources a new repertoire had to be produced which, as indicated previously, could be used by forwarding agents and customs staff in the many different countries rather than by chemists alone. The need was to create a listing of names which was both simple and as similar as possible in each of the languages. This was not an easy task because of a number of problems.

(1) There are a number of reasonably well developed systems for establishing chemical names in certain languages. English is well represented in these systems but the coverage of other languages varies considerably. The coverage given by the rules of the IUPAC is an example of an existing system of this type.

(2) The existing systematic listings are often too specialised and too technical for customs/commercial use. Names used in the preparation of Chemical Abstracts (CA) demonstrate this problem.

(3) The same chemical products often have different common names in the various languages complicating the problems of translation. In addition different countries have adopted differing conventions for specifying chemical names. Some of the differences are purely linguistic whilst others stem from variations in approach.

(4) The size and complexity of the task necessitates the use of computers and this introduces further constraints. For example, Greek characters, upper and

lower case characters, accents, italic characters and various brackets could not be used in the production of the repertoire.

It was therefore necessary to consider other solutions which were not always completely satisfactory.

(d) SOLUTIONS ADOPTED IN DEVELOPING A REPERTOIRE

The lack of comprehensive internationally agreed rules to meet the requirements has meant that problems have to be solved in the best possible manner. The following concepts were adopted in preparing the data base.

For all products which were the subject of an International Organization for Standardization (ISO) denomination, the ISO denomination was used as the preferred name. This covered particularly pesticides and phytopharmaceutical products. In the absence of an ISO denomination, the International Non-proprietary Names (INN) approved by the World Health Organization (WHO) were taken. This covered particularly pharmaceutical products. For certain derived substances such as salts and esters an INN (Modified), INNM, was used.

The ISO, INN and INNM denominations had the advantage of being short, and similar in English, French, as well as in Spanish for the INN. This similarity helped greatly in producing denominations in Italian, Danish, German and Dutch which had not previously existed.

Where no ISO, INN or INNM denominations were found, the IUPAC nomenclature and rules were used. These were not always entirely satisfactory because (i) systematic IUPAC names tend to be somewhat long and complicated for commercial/customs use, (ii) IUPAC names and rules often do not exist in languages other than English, and (iii) the IUPAC rules allow in many cases a certain number of options. For example, in the English version of the IUPAC rules the position number can appear either before the related name or just before the chemical function. The French version of the IUPAC rules is different since it uses locants after the function in the case of numbers, or before the function in the case of letters. Where options were available the most widely used version was chosen. In an attempt to harmonise and simplify, the position of the numbers has been made the same in all languages – in front of the function to which they refer.

There are no general IUPAC rules for some languages such as Danish and Italian and chemical descriptions were created based upon the nearest related tongue – English for the Danish version, French for the Italian. In certain areas the individual characteristics of the language had to be retained for clarity.

(e) THE CONTENT OF THE REPERTOIRE

Work on the production of the multilingual repertoire was possible once the rules had been determined and certain technical problems on the use of computers had been solved. The final repertoire consists of some 20,000 chemical denominations, involving 16,000 preferred names, and 4,000 synonyms. It contains all chemical products commonly traded internationally indicating their correct classification in the EC *Common Customs Tariff* and providing the basis for listing of the Community's trade statistics codes. The repertoire provides a computerised chemical data base which can be linked with other data bases and exploited for other purposes as necessary. The repertoire data base provides input to the environmental chemicals data base (ECDIN). The basic listing has been developed in Danish, Dutch, English, French, German and Italian and a Spanish version is in preparation (1981).

Each of the 140,000 names has been individually studied to ensure that it is up-to-date and correct. In a high proportion of cases, names have been created in languages where they did not already exist. The names have been transcribed into the necessary computer input medium, placed on a computer file and then checked and corrected as necessary. Much of the work on the English listing was done by staff at LGC whilst a team of dedicated chemists/linguists worked in Brussels on other language versions.

The layout of the multilingual version of the repertoire is as follows:

1. A numerical reference.
2. The classification of the product in the *Common Customs Tariff.*
3. The denomination in each of the Community languages preceded by the language symbol:

   DA, Danish; DE, German; EN, English; FR, French; IT, Italian; NE, Dutch.
4. The chemical name.

There are six separate volumes in addition to the multilingual version, each giving an alphabetic listing of chemicals in each of the Community languages. In these lists the products are given in alphabetical order and are preceded by the reference number and tariff classification. Once a chemical name has been found in any of the alphabetic versions, the reference number enables the user to identify all chemical names used for the product in question, including possible synonyms, in each of the Community languages by the use of the multilanguage version.

The data base has been prepared in a manner that allows provision for the inclusion of the NIMEXE (Statistical) code and Chemical Abstracts Service (CAS) Registry Numbers. It has been designed so that amendments can be readily introduced, synonyms found and full or partial listings in alphabetical numerical or tariff number order can be produced.

(f) FUTURE DEVELOPMENTS

The repertoire will need continuous amendment to ensure that it is kept up to date and any errors are eliminated. This task will be aided by the feedback from the national customs administrations and CEFIC. Comments received from individual users will also be useful.

Further exploitation of the data base is being examined. This will include a consideration of whether or not to publish CA references, and whether the data base should be made available 'on-line' to users through the EURONET system. Publication of a companion listing showing the NIMEXE (Statistical) codes is also expected.

The EC is expected to adopt a new integrated customs tariff and statistical nomenclature based upon the so-called 'Harmonised System for the Description and Coding of Goods' in about 1985. This system is under development in the Customs Cooperation Council. It brings together the customs tariff and statistical requirements and identifies them by a single code. The changes will be very significant but fortunately, so far as the repertoire is concerned, it is estimated that about 70 per cent of the work will be undertaken automatically using computerised modification procedures.

(g) THE NEED FOR A CHANGE IN ATTITUDE

International trade is essential to the future well-being of Europe and to other countries in the world. Chemical products represent a large and important sector of international trade. For trade, customs and statistics purposes there is a need to be able to identify these products by names which are relatively simple. Long, complicated chemical denominations of formulae cannot be coped with. Because English is used to such a great extent as a means of communication in the scientific field there is a tendency to forget the existence of other languages. English is not accepted as a unique language in commercial circles and therefore it is necessary to develop systems which facilitate international trade. Not only commerce but also scientists would benefit if a simpler and more coherent international nomenclature for chemical products could be produced. The aim should be to simplify the denominations but also, so far as is practicable, harmonise the word roots, prefixes and suffixes to be used, as well as the place in which they appear in the denomination.

The EC has gone some way in this direction by the development of a multilingual repertoire. In many cases it has had to invent its own standards although all concerned with the project would have preferred to adopt guidelines which had already been agreed at international level. Relevant bodies, particularly WHO, ISO and IUPAC, should recognise the need for simplifed and harmonised international designations for those chemical products which are commonly traded internationally and put their efforts into developing nomenclature which can be readily used in commerce.

CHAPTER 11

Biochemical nomenclature

Dr H. B. F. DIXON
Secretary, IUB-IUPAC Joint Commission on Biochemical Nomenclature
(University of Cambridge, United Kingdom)

ORGANISATION

To try to deal with problems of biochemical nomenclature, the International Union of Pure and Applied Chemistry (IUPAC) and the International Union of Biochemistry (IUB) have set up two nomenclature committees, the IUB–IUPAC Joint Commission on Biochemical Nomenclature (JCBN) and the Nomenclature Committee of IUB (NC–IUB). I write from the viewpoint of these committees.

Problems come to us in various ways. We may ourselves foresee problems, but more often others put them forward. We usually refer them to expert panels, set up under a convener. This is because the two small committees do not have the range of expertise to deal with problems in detail. When the experts report, we look carefully at what they suggest, and modify it, if necessary, for the general user. Communication between the experts in a single subject is not much of a problem; it is to people outside the small circle of experts where communication is harder, so the general assessor has an important part in framing recommendations on nomenclature.

CONSULTATION

When we have agreed with an expert panel on preliminary recommendations, we canvass a wider variety of opinion. In this task we are immensely helped by the IUB body, the Committee of Editors of Biochemical Journals (CEBJ) (see Table 1). The members are the eleven main general journals of biochemistry in the world. There are also 53 corresponding members, some of which are journals like *Nature* and *Science*, which publish biochemistry amongst many other subjects, others are journals that publish papers on specialist parts of biochemistry, and others are reviewing journals. This body helps by sending out draft recommendations to those who are editing and provides a wide variety of opinion before the drafts are modified and confirmed. CEBJ also helps by suggesting members for the expert panels.

Table 1 – Composition of the IUB Committee of Editors of Biochemical Journals

Members (11)
Arch. Biochem. Biophys.
Biochem. J.
Biochemistry
Biochim. Biophys. Acta
Biochimie
Biokhimiya
Eur. J. Biochem.
Hoppe-Seyler's Z. Physiol. Chem.
J. Biochem. Tokyo
J. Biol. Chem.
J. Mol. Biol.
Corresponding Members (53)

ENZYME CLASSIFICATION

The work of NC-IUB is varied. One of its tasks is to produce and revise *Enzyme Nomenclature*, the list of enzymes, which is useful to biochemists as an up-to-date list of all reported enzymes. One of the reasons NC-IUB exists as an entity distinct from JCBN is to allow those aspects that concern IUB alone to be separate from IUPAC. Each publication of IUB is put out to tender. At the time of writing *Enzyme Nomenclature* is published by Academic Press and lists over two thousand enzymes. Enzymes in this list are classified by the reaction they catalyse and are contained in six classes (Table 2).

The first three classes have names that are self-explanatory. Lyases are enzymes whose reaction is to separate a single substance into two. 'Ligases' is rather a local word; their reaction is a condensation with a concomitant hydrolysis of a nucleoside triphosphate, a very common reaction. These classes are broken down into sub-classes and sub-sub-classes in the way shown. Class 1 means oxidoreductases; 1.1 contains any enzyme that interconverts an alcohol and a carbonyl compound; 1.1.1 implies a particular hydrogen acceptor. There are in fact 177 members of the 1.1.1 group, from 1.1.1.1, an alcohol dehydrogenase that turns out to be rather unspecific, to 1.1.1.177, which does a biochemically rather unusual dehydrogenation on C-1 of glycerol 3-phosphate rather than the usual dehydrogenation at C-2.

Table 2 – Classification of enzymes in *Enzyme Nomenclature*

Classes of enzymes

1. Oxidoreductases
2. Transferases
3. Hydrolases
4. Lyases — These catalyse reactions of the type A = B + C
5. Isomerases
6. Ligases

Examples of subdivisions from class 1

1.1	$>CH-OH \rightleftharpoons >C=O$
1.1.1	NAD^+ or $NADP^+$ is the hydrogen acceptor

Two individual enzymes of sub-sub-class 1.1.1 are:

1.1.1.1	Alcohol dehydrogenase
1.1.1.177	Glycerol-3-phosphate 1-dehydrogenase

The first edition of this list was published by the Enzyme Commission. It is therefore usual when people cite these numbers in order to specify an enzyme to prefix its numbers with the lette s EC, e.g. EC 1.1.1.177. The alternative meaning of EC has resulted in some confusion: economists have used computer retrieval techniques to recover some European Community document only to find an entry on an enzyme!

There are more serious snags. All enzymes are listed by reaction, and this can give problems, as with adenosinetriphosphatase. This is an enzyme that catalyses the reaction of ATP plus water to ADP and orthophosphate. The action of muscle myosin and actin is one example, i.e. the system responsible for mechanical work by muscle. Another is the sodium–potassium pump, which pumps sodium out of the body cells into the extracellular fluid at a rate of some hundreds of amperes in the human body. The calcium pump is yet another: every time a muscle relaxes, it has to pump away the calcium that was activating muscle action. A final example is the pump that runs backwards to make the ATP used in the foregoing reactions; the ion gradient it uses is the proton-gradient across the mitochondrial membranes. All these utterly different proteins are a single entry in this list, and this is unsatisfactory. We may be able to distinguish between some of them by regarding sodium in one compartment as a reactant and sodium in another compartment as a product. But this changes the present basis of classification, which NC-IUB is hesitant to do.

ENZYME ACTIVITY

Another action of NC-IUB has been to define enzyme activity (*Eur. J. Biochem.*, 1979, **97**, 319). We need to measure the rate of conversion catalysed under specified conditions. That will usually have units such as moles per second, and this quantity is a measure of the amount of enzyme. NC-IUB thought it useful to give a special name to the unit 'mole per second', and suggested 'katal'. This has not found wide acceptance, possibly because it was launched rather badly. The advantage is the same as the convenience of calling the reciprocal second a 'hertz' when measuring frequency, and a 'becquerel' when measuring radioactivity; a special name to indicate the context.

ENZYME KINETICS AND RATE OF REACTION

NC-IUB is about to publish recommendations on how to represent enzyme kinetics. In this work we have been helped by a recent change in IUPAC recommendations (which we try to follow for chemical purposes). New IUPAC recommendations (*Pure Appl. Chem.*, 1981, **53**, 753) allow the use of the name 'rate of reaction' as already used by all biochemists and probably by all solution chemists, for the intensive quantity of amount of substance per unit volume per unit time (see Table 3). Gas kineticists do not find this quantity convenient because their volumes may change, so they use the quantity rate of conversion, i.e. the extensive quantity whose units are mol s^{-1}, which was previously called 'rate of reaction' under older IUPAC recommendations.

Table 3 – Quantities connected with rates

Quantity	Symbol	Dimensions	Units
Rate of reaction	v	$n\,l^{-3}\,t^{-1}$	mol dm^{-3} s^{-1}, mol l^{-1} s^{-1}, M s^{-1}
Rate of conversion	$\frac{d\xi}{dt}$, $\dot{\xi}$	$n\,t^{-1}$	mol s^{-1}

ENZYME KINETICS AND THE DESIGNATION OF RATE CONSTANTS

Another problem in expressing any kind of kinetics, chemical or biochemical, is how to designate rate constants. IUPAC recommendations are shown in Figure 1(a). The rate constant for the forward and reverse reactions of a single step must be easily identified and they are given positive and negative subscripts. Nevertheless biochemists have used very different systems. Some object to the system of Figure 1(a) because their computer languages cannot handle negative subscripts. Both the systems of Figures 1(b) and 1(c) have been widely used. The expert kineticist may feel no need for new recommendations as he can handle many inconsistent systems. Nevertheless the first-year undergraduate or the worker in a neighbouring discipline who wants to look up just a little about enzyme kinetics has other needs. He would benefit if all presentations used the same symbols. The NC–IUB compromise is to prefer the system of Figure 1(a) but not discourage the others, except to request those who use them to say exactly which system they are using. We hope to avoid constraining people to use systems that cope only with present knowledge and to acknowledge that a different system may be needed in the future.

$$\text{(a)}\quad A \underset{k_{-1}}{\overset{k_1}{\rightleftharpoons}} B \underset{k_{-2}}{\overset{k_2}{\rightleftharpoons}} C \underset{k_{-3}}{\overset{k_3}{\rightleftharpoons}} D$$

$$\text{(b)}\quad A \underset{k_2}{\overset{k_1}{\rightleftharpoons}} B \underset{k_4}{\overset{k_3}{\rightleftharpoons}} C \underset{k_6}{\overset{k_5}{\rightleftharpoons}} D$$

$$\text{(c)}\quad \overset{1}{A} \underset{k_{21}}{\overset{k_{12}}{\rightleftharpoons}} \overset{2}{B} \underset{k_{32}}{\overset{k_{23}}{\rightleftharpoons}} \overset{3}{C} \underset{k_{43}}{\overset{k_{34}}{\rightleftharpoons}} \overset{4}{D}$$

Figure 1 – Methods of distinguishing rate constants.
Method (a) is the one recommended by IUPAC.

JCBN AND ITS TASKS

A lot of JCBN's work is concerned with naming compounds, e.g. in the steriod rules, the carbohydrate rules, the rules for phosphorus-containing compounds. The Biochemical Society publishes for IUB a compendium, *Biochemical Nomenclature and Related Documents* (1978). It contains the recommendations made by CBN, the predecessor of JCBN and NC–IUB, and also relevant sections of the IUPAC *Nomenclature of Organic Chemistry,* namely Section E on stereochemistry, Section F for natural compounds, and Section H for isotopic replacement. Section F particularly gives advice on how to give a trivial name to a new compound without confusing other workers by implying unintended meanings. The compendium does not contain documents from NC–IUB and JCBN since they have only been in existence since 1977. Several more recent documents on nomenclature have been published, e.g. on unsaturated and branched-chain sugars, and we hope to continue with recommendations on tocopherols and vitamin D. Summaries of these recommendations are attached as an appendix to this chapter.

SPECIFICATION OF CONFORMATION

Communication is not just naming compounds. Hence our documents include one (see Appendix) on naming conformations of five- and six-membered sugar rings. This had been published in a preliminary form by Dr Schwarz in *J. Chem.*

Soc.; it has now been slightly altered, and been approved by both IUPAC and IUB. The system is fairly straightforward: a conformation of β-D-glucose is shown in Figure 2. This conformation is called 'chair'; hence the *C*. A reference plane of the parallel opposite sides is chosen so that the lowest numbered carbon atom is not in that plane. Therefore the reference plane in Figure 2 is C-2, C-3, C-5, 0-5 to put the C-1 out of the plane. The ring is viewed from the side from which the numbering appears clockwise, i.e. from above in Figure 2. Hence C-4 is on the near side of that plane to the viewer, and C-1 is on the far side. Thus, '4' goes in front as a superscript and '1' goes behind as a subscript giving the designation $^4C_1$. Other proposals are in preparation on polynucleotide and polysaccharide conformation.

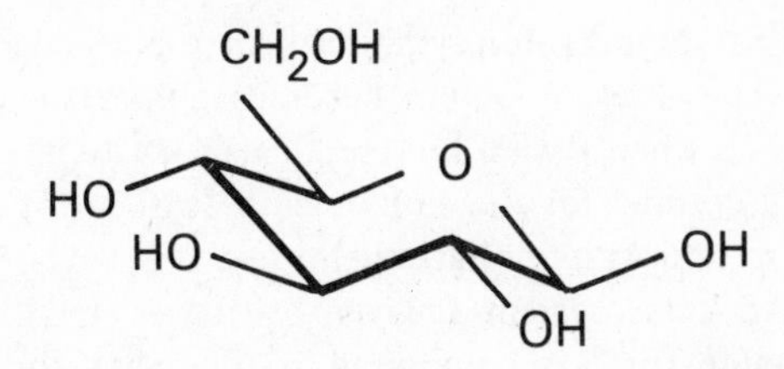

Figure 2 – β-D-Glucose in the $^4C_1$ conformation.

SYMBOLISM AND AMINO ACID RESIDUES

Symbolism seems at least as important as names of compounds for scientific communication. *The Guinness Book of Records* used to give the longest chemical name; it was a protein of just over 100 residues. No scientist would use such a name; he symbolises instead of naming. Much more so for the viral DNA whose sequence of 5,000 nucleotides has been found. The kind of symbolism used for amino acids is shown in Figure 3. The three letters, the first

Ala	signifies	NH_2-CHMe-COOH	and its ionised forms, i.e. the amino acid alanine
-Ala	signifies	-NH-CHMe-COOH	i.e. a C-terminal residue of alanine
Ala-	signifies	NH_2-CHMe-CO-	i.e. an N-terminal residue of alanine
-Ala-	signifies	-NH-CHMe-CO-	i.e. an alanine residue within a peptide or protein

Figure 3 – Features of the 3-letter system for amino acid residues.

of them upper case, mean the amino acid. That basic symbol is modified with hyphens. The hyphen on the left means removal of one H from the -NH_2; the hyphen on the right means removal of OH from the -COOH. The use of both hyphens indicates both removals, so that both hyphens are used for a residue in the middle of a protein and hence Ala-NH_2 means the amide of alanine. Some users find this hard to appreciate and think the NH_2 is the 2-amino group, which in fact is already included in the symbol 'Ala'. Substitution in the side chain (Figure 4) can also be indicated as shown for 3-chloroalanine.

$$\mathrm{NH{-}\overset{\displaystyle CH_2Cl}{\overset{|}{C}H}{-}CO{-}} \text{ is represented as } -\overset{\displaystyle Cl}{\overset{|}{\mathrm{Ala}}}- \text{ or } -\mathrm{Ala(Cl)}-$$

Figure 4 – Representation of side-chain substitutions.

Another local set of rules for symbolism in a particular field is used for longer sequences. The 3-letter system becomes cumbersome when there are about 100 residues or when several long sequences are to be compared and so a one-letter code is also used for the amino acids. It does not cope so well with modified residues, so both systems have their uses.

A use of the 3-letter system is given by the compound Tos-Phe-CH_2Cl (Figure 5). Any name for this compound is long, but the symbol gives its structure clearly. Although regarded as containing a residue of phenylalanine, it is not a carboxylic acid derivative in the normal sense. The enzyme chymotrypsin

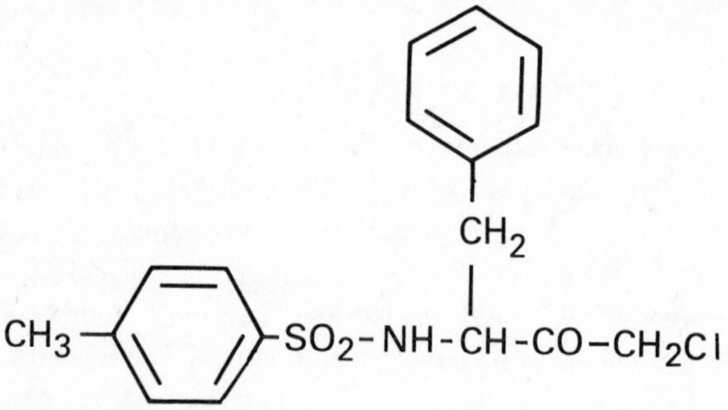

CH_2

CH_3–C_6H_4–SO_2-NH-CH-CO–CH_2Cl

Figure 5 – The structure of Tos-Phe-CH_2Cl, an inhibitor of chymotrypsin.

is too stupid to spot that, and so binds this compound at its active site, and a nucleophilic group of the enzyme attacks and knocks out the Cl^-. Hence the rest of the substance is bound and the enzyme activity is destroyed. The compound is quite important as a proteinase inhibitor, and a user might need to mention it in the titles of papers, but a chemical name or a formula really would not fit. The symbol Tos-Phe-CH_2Cl conveniently conveys the structure without too many other definitions.

We have been asked about the alternative symbols for 1-fluoro-2,4 dinitrobenzene, FDNB and Dnp-F. This is the compound with which Sanger first found protein sequences by labelling the amino groups that were free in protein (Figure 6) and he called it by the initials FDNB. The trouble with

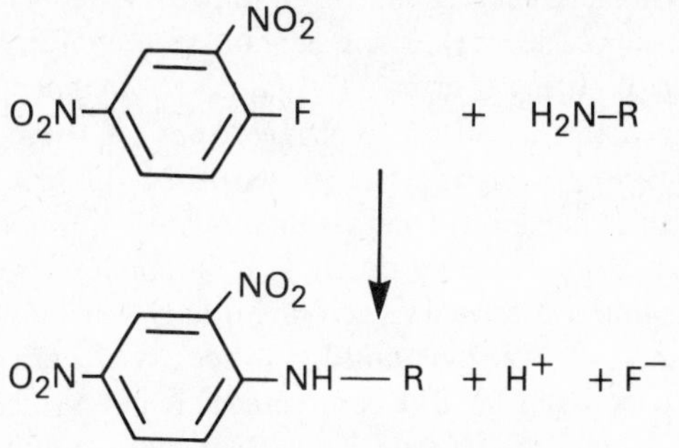

(a) FDNB + NH_2-R → Dnp-NH-R + H^+ + F^-

(b) Dnp-F + NH_2-R → Dnp-NH-R + H^+ + F^-

Figure 6 – Dinitrophenylation of an amine, and its representation with two different symbols for the reagent.

using initials is that FDNB in an equation (Figure 6(a)) makes that equation incomprehensible. A second symbol was introduced, a symbol for local use, namely Dnp-F. It allows the writing of a much more logical equation (b) because the Dnp-group keeps its identity in the course of reaction. This new symbol appears in the JCBN recommendations on peptide symbolism.

SYMBOLISM AND OLIGOSACCHARIDES

Approval has now been given by IUPAC and IUB on symbolism of oligosaccharide chains. Figure 7 shows a structure and the ways of symbolising it.

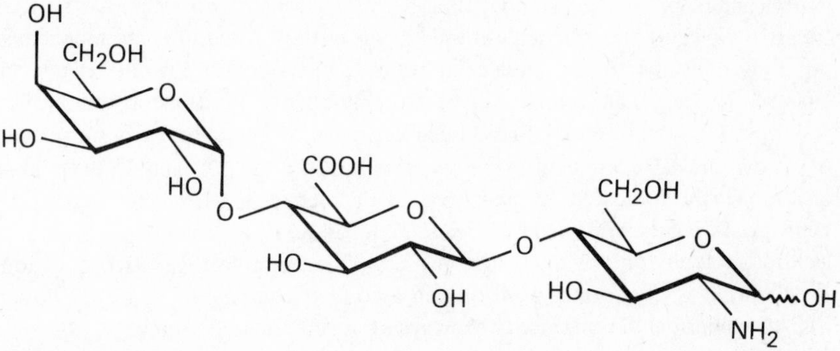

Extended representation: α-D-Gal*p*-(1→4)-β-D-Glc*p*A-(1→4)-D-Glc*p*N

Condensed representation: Gal(α1-4)GlcA(β1-4)GlcN

Figure 7 – Methods of representing oligosaccharide structure.

The first has been used in the literature for a very long time and the only change has been to codify it. The residue on the left is galactose. It is written with the italic *p* to show that it is a pyranose form. It is linked by its C–1 to a glucose unit that has been modified by the A to form glucuronic acid, i.e. C–6 is oxidised. The right-hand glucose residue is modified by the N to be glucosamine (strictly 2-amino-2-deoxyglucose). There is another system in the literature and it has the advantage of being shorter. NC–IUB have recommended both systems to overcome the need to force users to change from a satisfactory system. The designers of the shorter system argued that, whereas D-glucose and D-galactose are common, and the L-sugars are almost unknown, it was a waste of space to specify D for each residue. Similarly pyranose is far commoner than furanose, so by leaving out the symbol *p* as well (specifying it only when unusual) the system becomes more concise. It would be possible to specify an even shorter system if there had been no existing usage. The compounds are polysaccharides; they have got to be capable of hydrolysis, and hence they must have glycoside bonds. Hence the '1' numerals are redundant, because, for example, galactose must be joined through its C–1 as it is an aldose. These redundant numbers are given to avoid a change from accepted practice.

MOLECULAR MASS, MOLAR MASS AND RELATIVE MOLECULAR MASS

IUPAC recommendations are only useful in so far as they are applied. Biochemists, like most chemists, often think nomenclature a bore and do not want to know what has been recommended. An example follows of an attempt being made to improve the usefulness and to bring together some recommendations that may have been going astray. The quantity once called 'molecular weight' is now 'relative molecular mass' in school teaching. The name 'molecular weight' is dimensionally wrong and confusing because a weight is a force. Nevertheless 'relative molecular mass' is seldom used in the current literature, and biochemists not only continue to use 'molecular weight', but symbolise it confusingly. An example is the gel electrophoresis of proteins under conditions where mobility is a function of molecular mass. Such gels have to be calibrated with proteins of known relative molecular masses, say 32,000, 24,000 and 15,000. These are frequently indicated in photographs of such gels with labels beside the appropriate bands, and the labels read: 32 K, 24 K, 15 K. The use of 'K' appears as though temperatures are being quoted. This example raises two questions. Firstly what is the correct position and secondly what should be done? Present IUPAC recommendations give three ways of saying the same thing:

the protein has a relative molecular mass (M_r) of 15,000;
the protein has a molar mass (M or M_m) of 15,000 g/mol;
the protein has a molecular mass of 15,000 u (atomic mass units).

'Relative molecular mass' is not popular both because it is cumbersome, and because, being a pure number, it has no unit and therefore 'kilo' and 'mega' cannot be used with it. Molar mass is fairly convenient but almost impossible to use adjectivally: a '15-kilogram-per-mole protein' does not trip off the tongue. Hence it would be useful to name the mass, except that the unit has both an awful name – atomic mass unit – and a crazy symbol – u. NC–IUB is therefore recommending a name quite widely used in the literature, and suggests that users should call the atomic mass unit a 'dalton' and symbolise it as Da. It then becomes easy to write in symbols 'a 44-kDa protein' and to pronounce in speech 'a 44-kilodalton protein'. Further the gel electrophoresis can be labelled with 30 kDa, 24 kDa, 15 kDa in a way that is easily understandable. To the purists it may be reprehensible to imply a quantity by the unit, but it should be clear that the quantities are masses and therefore molecular masses.

CONCLUSION

The range of the tasks undertaken by the nomenclature committees of IUPAC and IUB have been outlined in this chapter and problem areas identified to stimulate further discussion on outstanding areas where recommendations are needed.

APPENDIX

JCBN 1

Conformational Nomenclature for Five and Six-membered Ring Forms of Monosaccharides and their Derivatives (Recommendations 1980)

The system for designating six-membered ring-forms of monosaccharides as chair (C), boat (B), skew (S) and half-chair (H) forms is described, including the symbolism, e.g. $^{1}C_{4}$. In this symbolism a superscript before the letter means displacement of the atom indicated to the side from which the numbering of the ring appears clockwise, and a subscript after the symbol means displacement of the atom indicated to the opposite side. Five-membered rings are similarly described as envelope (E) and twist (T) forms. Several examples of conformations are illustrated.

Published: *Eur. J. Biochem.*, 1980, **111**, 295; *Pure Appl. Chem.*, 1981, **53**, 1901.

JCBN 2

Nomenclature of Unsaturated Monosaccarides (Recommendations 1980)

Recommendations are given for naming unsaturated monosaccarides. Double bonds are indicated with the infix 'en' and its use is described. Triple bonds and cumulative double bonds require 'deoxy' prefixes to convert sugars into derivatives on which the operator 'dehydro' can indicate double and triple bonds.

Published: *Eur. J. Biochem.*, 1981, **119**, 1; *Pure Appl. Chem.*, 1982, **54**, 207.

JCBN 3

Nomenclature of Branched-chain Monosaccharides (Recommendations 1980)
The naming of acyclic forms is described, and starts with rules for distinguishing the parent chain from the branches. This leads to naming the parent monosaccharide, including the naming of the *C*-substituent that forms the branch. Stereochemical conventions have to be extended to specify configuration: the two substitutents at one carbon on the parent chain are put in order of priority; that of higher priority is considered to replace hydroxyl and that of lower priority to replace hydrogen in assigning a name to the parent monosaccharide. Equivalent groups and terminal substitution are dealt with separately. Cyclic forms are then described. Many examples are given.

Published: *Eur. J. Biochem.*, 1981, **119**, 5; *Pure Appl. Chem.*, 1982, **54**, 211.

JCBN 4

Nomenclature of Tetrapyrroles (Recommendations 1978)

Published: *Pure Appl. Chem.*, 1979, **51**, 2251; *Eur. J. Biochem.*, 1980, **108**, 1.

Summarising report by P. Karlson: *Hoppe-Seyler's Z. Physiol. Chem.*, 1981, **362**, VII-XII.

JCBN 5

Nomenclature of Vitamin D (Recommendations 1981)
The recent revival of interest in vitamin D analogues has rendered cumbersome some of the older systems for naming these compounds. This document proposes an extended system of trivial and semisystematic names for the important metabolites of vitamin D.

The trivial names cholecalciferol and ergocalciferol included in the 1966 proposals on *Trivial Names of Miscellaneous Compounds of Importance in Biochemistry* are retained as alternatives to calciol and ercalciol respectively. The trivial names calcidiol, calcitriol and calcitetrol are reserved for the 3,25-diol, 1,3,25-triol and 1,3,24,25-tetrol respectively. It is hoped that these will replace undesirable abbreviations like 1α,25-$(OH)_2D_3$.

For semisystematic names the most important change from past practice is the recommendation that all stereochemistry in ring A or at C-6 or C-7 should use *R, S, E* or *Z*. This prevents confusion arising from the conformation of ring A relative to rings C and D where, for example, the steriod 3β-hydroxy group is frequently in the α region.

Published: *Eur. J. Biochem.*, 1982, **124**, 223; *Pure. Appl. Chem.*, 1982, **54**, 1511.

JCBN 6

Abbreviated Terminology of Oligosaccharide Chains (Recommendations 1980)
After a review of trivial and systematic names, symbols are given for specifying sugars, uronic acids, deoxysugars, aminodeoxysugars, anhydrosugars, etc. The

mode of linkage is then specified for unbranched and branched oligosaccharides, and the method of indicating substitution and branching is given. An appendix gives a condensed system of symbolism. Many examples are given, e.g. amygdalin is represented as β-D-Glc*p*-(1→6)-β-D-Glc*p*-O-CH(CN)Ph in the full system and as Glc(β1-6)Glc(β)-O-CH(CN)Ph in the condensed system.

Published: *J. Biol. Chem.*, 1982, **257**, 3347; *Pure Appl. Chem.*, 1982, **54**, 1517.

JCBN 7

Polysaccharide Nomenclature (Recommendations 1980)
Generic names such as polysaccharide are first defined, and then the use of the ending 'an' as in glucan, glucuronan, etc. The naming of polysaccharides containing aminodeoxysugar residues is described. Conventions are specified for heteropolysaccharides, which contain two or more kinds of sugar residue, so that the name can indicate features such as a backbone of a single type of residue. Methods of indicating substituents are also given.
Published: *J. Biol. Chem.*, 1982, **257**, 3352; *Pure Appl. Chem.*, 1982, **54**, 1523.

JCBN 8

Nomenclature of Retinoids (Recommendations 1981)
Previous proposals published in the 1966 *Trivial Names of Miscellaneous Compounds of Importance in Biochemistry* included the trivial names retinol, retinal and retinoic acid. These names are retained, but the numbering of the carbon skeleton is changed to conform to the 1975 *Nomenclature of Carotenoids*. Methods for naming metabolites and related molecules based on the stem 'retin-' are included. Because of the previous misleading use of the word retinene as a synonym for retinal the basic hydrocarbon also known as axerophthene is called deoxyretinol.

Published: *Eur. J. Biochem.*, 1982, **129**, 1.

JCBN 12

Nomenclature of Tocopherols and Related Compounds (Recommendations 1981)
Provisional recommendations published in 1974 on the nomenclature of tocopherols have been revised incorporating proposals agreed between the Commission of Biochemical Nomenclature (predecessor of JCBN) and the International Union of Nutritional Sciences. The only significant change from the previous version is to recommend that the old trivial names (*d*-α-tocopherol for *RRR*-α-tocopherol, 1-α-tocopherol for 2-epi-α-tocopherol, and *dl*-α-tocopherol sometimes used for 2-*ambo*-α-tocopherol and sometimes for *all-rac*-α-tocopherol) should no longer be used.

Published: *Eur. J. Biochem.*, 1982, **123**, 473; *Pure Appl. Chem.*, 1982, **54**, 1507.

JCBN 13

Symbols for Specifying the Conformation of Polysaccharide Chains

The notation for atomic numbering, interatomic distances, bond angles and ring shapes are described, and the rules for defining conformation of various side groups are discussed and illustrated. Residues are numbered from the reducing end, so that terminal transglycosylation does not alter the numbering of every residue in a chain. The glycosidic swivel joints are designated and the convention for torsion angles is given.

JCBN 14

Abbreviations and Symbols for Specifying the Conformation of Polynucleotide Chains

General principles of notation are first described, defining chain direction and the nucleotide unit, and specifying nomenclature for atom numbering, bond lengths, bond angles, torsion angles and conformational regions. The description of conformations for the bonds of the nucleotide unit is then given in detail for the sugar-phosphate backbone, the sugar ring, the N-glycosidic bond and the orientation of side groups. Recommendations are also made for the description of hydrogen bonds, base pair schemes and helical segments of polynucleotide chains. General examples of conformations are illustrated, The proposed nomenclature is consistent with that recommended for conformations of polypeptides and polysaccharides.

CHAPTER 12

The use of trivial and generic names

R. B. TRIGG
British Pharmacopoeia Commission, London, United Kingdom

For the purpose of modern commerce and scientific communication users may regard as acceptable those names, trivial, generic, non-proprietary approved, common – under whatever description – that have been carefully defined by an acknowledged public authority. 'Trivial' names approved by IUPAC which cannot be described as systematic in any way – acetic acid, aniline, toluene – can be regarded as acceptable for use but not those names handed down from antiquity, and in many cases of uncertain structure, that have not been enshrined by systematic nomenclature authorities. The difficulties which can arise from the use of ill-defined trivial names has been described by Egan and Godly.[1] The terms, 'approved' or 'common' can be restricted to names that have been artificially created by national and international non-proprietary nomenclature authorities to serve the needs of science and commerce by providing a short, distinctive, unique name to identify positively a chemical substance. The term 'generic' is used in this chapter to embrace both the antiquarian type of 'trivial' name and the artificial 'approved' or 'common' name. A generic name is a contradiction in terms – it is not generic but specific to a single chemical substance. It is, however, a simpler word to spell and pronounce than non-proprietary, two of the properties which are an advantage when choosing new generic names.

THE CURRENT POSITION

To the best of my knowledge, the adoption of generic names for new chemical substances intended for commercial use is centred on the guidelines provided by two international organizations, the World Health Organization (WHO) and the International Organization for Standardization (ISO), who between them have sought to bring rationality into what might otherwise be a chauvinistically chaotic situation. The pharmaceutical and pesticide industries between them have an enormous appetite for new substances, and WHO and ISO endeavour to maintain the semblance of order which now prevails in the commercially important matter of nomenclature. A list of authorities concerned with generic names is given in Table 1.

Table 1 – Generic name authorities

Pharmaceutical	Name designation
World Health Organization	INN
United States Adopted Names Council	USAN
British Pharmacopoeia Commission	BAN
French Pharmacopoeia Commission	BCF
Japanese Accepted Names Committee	JAN
Pesticidal	
International Organization for Standardization	
American National Standards Institute	
British Standards Institution	
French and German National Standards Organisations	
– all known as common names, ISO, ANSI, BSI, AFNOR, DIN respectively.	

Neither WHO nor ISO has any real authority with which to impose their recommendations. Both organisations require the co-operation of the relevant regulatory authority in each of their member states. In the field of pharmaceuticals in the United Kingdom, the British Pharmacopoeia Commission supported by staff of Medicines Division of the Department of Health and Social Security, is empowered under Section 100 of the Medicines Act 1968 and Statutory Instruments No. 1256 to select and publish new generic names for medicinal substances. These names are called British Approved Names (BAN), and the degree of close collaboration between the Commission and WHO is such that names for substances researched and developed abroad and published by WHO as International Non-proprietary Names (INN) are almost without exception adopted and approved for use, when required, in the United Kingdom. An occasional exception can arise when the name in question is in conflict with a trade mark already in use.

Similar collaboration exists between WHO and the national authorities of most of those countries supporting a research-based pharmaceutical industry. In the United States of America, it is the United States Adopted Names Council, a non-regulatory organisation in fact, whose nomenclature adoptions are given authority by the Food and Drugs Administration. In many other countries, as in the United Kingdom, it is the concern of the national pharmacopoeia authority to regulate the adoption of new pharmaceutical names.

In the pesticide sector, a parallel situation can be drawn. The names selected by ISO are to be published as an International Standard, but ISO member states will be required to enshrine the document as a national standard. Thus, in the United Kingdom, draft ISO Standard 1750 reflects British Standard BS:1831, and, although not fortified by an Act of Parliament, any pesticide manufacturer who departed from the recommendations of BS:1831 would be guilty of unprofessional practice. Almost all developed countries possess a national standards organisation to implement the recommendations of ISO.

It may be concluded that a one-way traffic in names exists between WHO, ISO and member states. This is not the case as WHO and ISO are simply offering the machinery by which international agreement on names can be achieved. In the first instance, it becomes the concern of the national authority in the country in which the new product has been researched to achieve a name acceptable to both the company in question, and itself. At this point the proposed name is fed into the WHO/ISO machinery by means of which it is circulated for comment to appointed experts in a representative selection of member states, or in the case of ISO, to national committees of participating member states. Either way, a unanimous or clear majority vote is taken as a signal that a name is suitable for international use. It would clearly be foolish at this level to insist on a unanimous vote since the vagaries of language are such that perhaps an occasional conflict with the trade name in Hungary or an undesirable connotation or difficulty of pronunciation in Japanese need not disqualify the proposed name from use in all other countries where its use is considered acceptable. The view then in both circles is that if a name is acceptable to a clear majority of member states, it should go forward for approval as an international recommendation.

At this point the paths of WHO and ISO divide. Selected WHO names are published every six months in a supplement to *WHO Chronicle,* WHO's journal, where they are known as Proposed INNs. The machinery allows for a four-month period of public comment, during which objections, usually based on commercial trade mark infringement, may be lodged, and if validated by the WHO panel of experts will be registered as formal objections. Such tarnished names will not proceed to become Recommended INNs, in which form the names are republished in *WHO Chronicle* some six to twelve months later. ISO, on the other hand, engages in a larger measure of pre-publication scrutiny. Public comment in the United Kingdom is allowed for by means of a proclamation in *BSI News,* the organ of the British Standards Institution which happens to hold the secretariatship of the ISO technical committee for pesticide nomenclature. In addition, the sponsoring manufacturer is required to submit the results of a search of the trade mark registries of the major manufacturing countries. Further hurdles to be cleared include preliminary enquiry followed later by letter ballot in the member states.

On balance, the WHO approach is probably the better. It allows the publication of about 150 new names each year, sometimes as little as six months from the date of the initial proposal from the manufacturer. The ISO publication machinery grinds more slowly, and several years may elapse before a new name finds itself in a Draft Addendum to ISO : 1750. Indeed the ISO standard itself is still only available in draft form. Final publication is awaited with increasing impatience.

GROUND RULES FOR SELECTION

Traditionally generic names have been devised, in the case of a natural product, from the botanical or zoological name of the species concerned; papaverine, penicillin and streptomycin are three very well known examples. With the dawn of the era of synthetic drugs it became the practice to coin generic names simply by stringing together key syllables taken from the systematic chemical name; inspection of any list of drug or pesticide names will instantly reveal numerous names simply composed of familiar elements of systematic radicals and roots. As rich veins of profitable research were discovered it then became the practice to hallmark each nugget with a unifying handle. Thus arose the family of sulphonamide anti-bacterials identified by the prefix 'sulpha-', the semi-synthetic penicillins each bearing the generic root '-cillin', and the phenothiazine tranquillisers many of which bear the root '-azine'. As more and more families of drug substances have been discovered during the last 30 years, so the practice of establishing a specific stem on which to graft any number of distinguishing prefixes has grown. The most important of these stems are listed and defined in most of the international and national drug name directories and a selection is given in Table 2.

It has, however, proved difficult to break away from the time-honoured practice of borrowing systematic terms to complete names devised using the prefix-stem approach. As manufacturers strive to find the ideal drug substance in any one family, it is inevitable that a series of substances very closely related structurally, finds its way to market. It is equally inevitable that they are distinguished in name simply by use of syllables indicative of the chemical differences; thus we have nitrazepam, flurazepam, flunitrazepam, bromazepam, etc. Clearly these same substitutionary radicals will serve to distinguish members of many different series, and the situation has long since reached the point when new names bearing initial syllables such as chloro-, meth-, phen-, etc. can no longer be entertained except in the most pressing of circumstances.

One of the two key principles underlying the selection of generic names in both the pharmaceutical and pesticide sector is that a name should be distinctive in sound and spelling. The continual use of syllables borrowed from systematic nomenclature is insidiously eroding this principle, and in order to preserve its value and to forestall those who criticise generic names on the grounds that they sound too alike, increasing use is now made of inert or abstract prefixes in conjunction with the generic stems. For example, in the well-known family of cardiac drugs based structurally on 2-aminopropanol and known generically as the -olol series, recent additions include nadolol and moprolol.

Table 2 – A selection of generic stems with examples

-azepam	diazepam group
-buzone	phenylbutazone group
-caine	local anaesthetics
cef-	cephalosporins
-cillin	penicillins
-cycline	tetracyclines
-metacin	indomethacin group
-olol	beta adrenergic blocking agents
-profen	ibuprofen group
prost	prostaglandins
sulfa-	sulphonamides

Stem	Example
-profen	anti-inflammatory propionic acids R–CH–COOH with CH_3 on the CH
-olol	beta adrenergic blocking agents X–OCH_2–CH–CH_2–NHY with OH on the CH

The value of the generic stem to the medical practitioner as an indicator of biological activity, pharmacological action, allergy potential or of the need for extra special attention to the monitoring of patient-response is such that the prefix-stem approach to drug nomenclature must be allowed to flourish, and it appears that this can only be achieved by a more flexible approach to the choice of initial syllables.

The value of the prefix-stem approach embraces the second major principle underlying the selection of a generic name, that is that the name chosen should reflect any biological or pharmacological similarity to any existing substance. It is surprising that there are some experts who firmly believe that every generic name should be as unique as a trade mark, and for this reason it has not yet been possible to embody this principle within the rules of guidance in the selection of names for pesticides.

CHOICE AND AVAILABILITY

Each and every new chemical substance synthesised for an examination of its commercial potential enjoys a series of identities usually beginning with a laboratory code number consisting of the initials of the chemist and a sequential serial number. Any idiosyncracies of worker or compound may well be reflected in the adoption of a colloquial name such as Smith's acid or Brown's ketone. In 1970 I made an attempt to prepare the acid chloride of *o*-phenanthridinylbenzoic acid by treating it with thionyl chloride. An innocuous white powder was expected but instead a brilliant red, highly reactive product was obtained whose structure was never identified. This became known affectionately throughout the laboratory as Trigg's Red. Code numbers and trivial names of this kind adequately serve the internal needs of laboratory workers, but later have to be abandoned in the interests of sensible communication. In most cases the task of devising a non-proprietary name falls to a research committee that may only bring as much wit to the task as the group that set out to design a horse and came up with a camel.

Choosing a name requires the most careful attention. Even if the fundamental principle of analogy is carefully followed, insufficient consideration is given to the important matter of availability.

No name however euphonious and scientifically appropriate can be entertained if it fails to fulfil the requirement of distinctiveness. The criteria for this are threefold:

(1) the name should be distinctive in sound and spelling;
(2) the name should not be inconveniently long;
(3) the name should be free from confusion with existing generic names and free from commercial conflict with trade marks.

Each of these criteria is of course entirely subjective. Critics of generic prescribing in medicine, for example, will produce examples of confusing generic names, chlorpropamide and chlorpromazine, clotrimazole and cotrimoxazole, in response to which the defender can recite pairs of confusing trade names with equal facility. These unfortunate similarities cannot be denied. Finding distinctive names becomes correspondingly more difficult as the list of new materials becomes longer. The key to success lies ideally in the use of computer-based directories in which not only the names themselves are filed alphabetically, but also each name is filed letter-by-letter. For example, if the name promethazine is filed in the following way:

*p*romethazine	
*r*omethazine	p
*o*methazine	pr
*m*ethazine	pro
*e*thazine	prom
*t*hazine	prome
*h*azine	promet
*a*zine	prometh
*z*ine	prometha
*i*ne	promethaz

and all the names on the file are treated similarly, it will be seen that all names containing the common suffix '-azine' will appear alongside each other. By the same token any other name containing the sequence '-methazine' will immediately show up. The value of file of this kind as an aid to eliminating misleading connotations and confusingly similar names in the choice of a new generic name should be immediately apparent.

A particular difficulty in screening proposed names for conflict with both other generic names and trade marks concerns the detection of names with a similar sequence of vowels but containing a different set of consonants. Such conflicts can be very real but are extremely difficult to detect simply by manually searching an index. However, if a computer index can be programmed to file names classified by vowel sequence, a very powerful search tool becomes available. For example, if the name pr*o*m*e*th*a*z*i*ne is taken once again then a vowel-sequence print-out of the INN file would reveal the following set of names, each containing the sequence o-e-a-i: (the terminal 'e' is ignored).

promethazine
ortetamine
chlorprenaline
cloperastine
profenamine
proheptazine
ftormethazine

In fact, none of these names could be held to be in conflict with any of the others, but supposing the name *dronefasine* existed. Few could deny a conflict in speech with promethazine, but it is likely that few would have found it using traditional search methods.

WHO has set up computer files of these kinds to facilitate the work of selection of INNs. Unfortunately the computer print-outs, by means of which the files are most conveniently used off-line, are not commercially available, but copies of the master tape and the associated software can be made available to recognised authorities. This is useful but the real need is for a comprehensive computer file of all official generic names, recognised trivial names and trade marks registered in Class 5. A file of this type would serve the interests of both the pharmaceutical and agrochemical industries.

Scientists should avoid the precipitate use of a newly chosen generic name in research papers. The pharmaceutical literature is littered with the corpses of prematurely used names which subsequently were rejected by a national nomenclature authority for reasons of non-availability or failure to observe the basic principles. Perhaps the best-known example is the name 'prostacyclin' which has appeared in many medical journals in recent years, and has even featured in articles in the lay press proclaiming its exciting potential for the treatment of blood platelet aggregation. In this instance the manufacturer overlooked the fact that the stem '-cycline' is to medical workers synonymous with substances in the tetracycline antibiotic family. Furthermore it took no account of a Hoechst trade-mark 'HOSTACYCLINE'. Accordingly prostacyclin has been rejected by the major nomenclature agencies in favour of the name epoprostenol which, while not being anything like such a magnetic name as prostacyclin, at least more accurately reflects the structural configuration of the compound in question.

APPLICATION

Protection of a name is a question of the greatest concern in both the generic name and trade-mark spheres. Both pharmaceutical and pesticide name applications in the United Kingdom are processed by expert committees composed of representatives from medicine, pharmacy, government service and manufacturers' associations. A generic name serves to identify uniquely a chemical substance by means of a short, simple name, available for anyone to use. It does not serve to identify uniquely a single product which is the purpose of a brand name. It is of paramount importance that the two ideas are not confused. The brand name or trade-mark serves to identify the source of the product and the generic name to identify the nature of the product – two names each with a different purpose.

A traditional view, still widely held in some quarters, is that manufacturers are little concerned about generic names and deliberately contrive to secure long and unwieldy names to ensure the preferred use of their trade-marks, particularly in regard to prescribing. The generic authorities are alert to this practice and nowadays rarely entertain a generic proposal in excess of five syllables. Understandably, the manufacturer's concern is to sell his product at the expense of rival brands of products containing the same ingredient. This can only be achieved through the use of that manufacturer's trade-mark which brings to bear the goodwill, quality assurance and reputation that the company hopes it enjoys with medicinal products. The question that remains is what incentive or compulsion is there for the manufacturer to actually use the generic name? The argument in 'the brand name/generic name prescribing controversy' claim that if drugs were prescribed by generic name, the dispensing pharmacist would be free to dispense whatever in his professional judgement was the most cost-effective brand, assuming more than one brand existed. If, in exercising that judgement, pharmacists chose to dispense what are known as unbranded generic equivalents, manufactured by companies without research and development obligations, the possible saving to the United Kingdom National Health Service on the 13 most-widely prescribed medicines has been estimated at £25m. On the other hand, if a prescription specifies a brand name, the pharmacist's legal obligation is to supply that particular product whatever the cost. The defence of this approach is two-fold. The branded product enjoys the reputation of the manufacturer for quality and consistency of presentation, although this claim has been found wanting on more than one occasion; but, and perhaps this second point is more defensible, research-based industry needs to be able to recoup the enormous costs of research and development and to ensure a future supply of new products. Quality of product, however, is not left to the whim of the manufacturer but is imposed jointly by the requirements of national pharmacopoeia and licensing authorities. In the United Kingdom, the British Pharmacopoeia (BP) lays down purity specifications for some 2,000 ingredients and their retail preparations. These specifications, or monographs as they are known, make use of generic names. So, if a manufacturer wishes to sell a product for which the BP provides a shelf-life specification, he is obliged under the Medicines Act 1968 to comply with that specification. It is clearly in his own interest to label his product in such a way that it is clearly seen that, for example, Mogadon Tablets comply with BP requirements for nitrazepam tablets. Indeed, the Medicines (Labelling) Regulations 1976 insist upon such a procedure. Section 3 of these regulations requires that each and every pharmaceutical product offered for sale in the United Kingdom is clearly labelled with the appropriate non-proprietary name, which is defined elsewhere in the regulations as the BP monograph title, approved name, INN, and, as a final catch-all, 'any other scientific name descriptive of the true nature of the medicinal product or ingredient'. The regulations also require that if the INN is different from the monograph title or approved name then it too must appear on the label. Logistics therefore demand that generic names be kept short and simple.

Similarly, in the agrochemical sector, the UK Pesticides Safety Precautions Scheme (Revised 1979), Appendix E, paragraph 4.1.2 requires notification of 'the common name(s) of the active ingredient(s) (according to BSI) or, if a common name is not available, the chemical name according to IUPAC Rules as interpreted by the Chemical Society'. For most chemicals this is a statutory requirement under the Farm and Garden Chemicals Regulations 1971.

REFERENCE

1. H. Egan and E.W. Godly, *Chem. Brit.*, 1980, **16**, 16.

CHAPTER 13

Some legal implications with trade marks and common names

D. ROSSITTER
Lilly Industries Ltd, London, United Kingdom

Derek Rossitter is, so I believe, a human being but, happily for humanity, not all human beings are Derek Rossitter. The distinction is between a particular member of a species and the species concerned. If one is ever to begin to understand the different legal implications arising from use of trade marks on the one hand (which conceptually designate particularised commercial origin) and the so called 'generic names' on the other hand, it is essential to grasp this elementary distinction.

Example: *Trade mark* – KEFLEX *Generic* – cephalexin, itself a member of the cephalosporin series

The controversy concerning the advisability or otherwise of using trade marks for preparations of the substances is outside the scope of this chapter. There are, however, two quite different types of nomenclature – a private and a public system – having quite different purposes and therefore it necessarily follows that their construction, protection and legal implications differ fundamentally.

A trade mark is intended to distinguish one person's goods from another's. It is private property (e.g. KEFLEX belongs to Eli Lilly & Co., USA), a species of property referred to as 'intellectual property' in common with patents and copyright. As such it is entitled to the type of protection commonly associated with private property with special additional protection designed, in essence, to prevent deception. Because it is private property the onus for its protection falls primarily on the trade mark's proprietor just as the onus for protecting a gold watch falls on its owner, although the owner is entitled to the protection of the law if somebody attempts illegally or unjustly to take it away.

The generic name is intended to provide a means by which a substance can be described easily without recourse to its systematic name (assuming that the full chemical name is too complex for convenient use). Obviously, the principal use to which such a name is put is in science. Since science is essentially international in character it is important, therefore, that the name so designed should be acceptable everywhere without regard to political, legal or linguistic barriers.

Trademarks, being private property, are subject to the national laws of the countries in which they exist. Most countries have laws which are specifically referred to as the 'Trade Mark Law' or some such title (e.g. in the United Kingdom the relevant law is the Trade Marks Act 1938) but it is very important to appreciate that such laws are only part of the total law – which, of course, includes obligations under international treaties as well as such obviously relevant law as that relating to false descriptions, public safety, sale of goods, unfair trading and the like.

Since trade marks distinguish a particular trader's goods from those of other traders and since trade may be simply local or very much international depending on the size of the trading concern and the nature of its business it necessarily follows that trade mark ownership may exist on a very limited local scale or on a very wide international basis. Furthermore, the nature of the goods to which the trade marks may be applied may theoretically cover every conceivable kind of object and, in a more specialised sense, activity – these latter being known as service marks.

Merely because a company is small, or has a limited area of trading, or is dealing in a very specialised type of product, does not mean that its trade mark is unimportant to it – in fact the converse is true. The trade mark DISTALGESIC, for example, is of immense importance in the United Kingdom and some few other places, but virtually unknown elsewhere.

Trade mark protection arises in two ways, sometimes separable and sometimes inseparable. Under the first method it arises by acceptance for entry upon an official register. Under the second method it arises by reason of proven use. In some jurisdictions mere registration confers rights; in others registration is dependent upon use. The situation is extremely complex. In addition to this, in some jurisdictions no really effective official control is exercised over what is or is not able to be entered on the register, disputes being left to be settled by private litigation or agreement. In others, entry to the register can only be gained after a searching official examination by the authorities in regard to, for example, the existence of similar previous names for similar goods, misleading connotations in the proposed name, or similarities with established common names. In such cases, however, the position may be undermined by the fact, and such is the case in the United Kingdom, that since rights can be established by use alone without registration, the registration process can be ignored or by-passed. However, to do so has some substantial disadvantages.

From the preceding it can be readily seen that an assessment of apparently conflicting rights arising under trade mark legislation throughout the world (and a generic name is after all supposed to be the same everywhere) is a complex and highly skilled task involving great knowledge, not only of different systems of law but also of the methods of establishing use and of the nature of the goods involved.

It must be appreciated that although, superficially, certain categories of goods may appear to be quite different, when dealing with the chemical substances to which generic names are applied (and indeed with the branded preparations themselves) these distinctions can become quite hazy. For example the following fields obviously gradually merge into each other:

General chemicals
Cosmetic substances
Pharmaceutical preparations for human use
Veterinary preparations
Medicated feed additives
Ordinary foodstuffs and drinks
Diagnostics
Herbicides and pesticides and similar goods

This covers a very wide area of science and commerce and it must be obvious that, given the right circumstances, confusion between any two such areas could lead to disaster. Consider, for example, rodenticides and foods having similar names. The question is: how can such confusion and danger be reasonably avoided?

So far as trade mark rights are concerned, since these are private property it might reasonably be argued that, given perhaps some judicious governmental intervention, it is up to the owners themselves to fight their own battles and make sure that their product names do not land them in some situation which may damage their rights or even tarnish their reputation.

Some authorities take stronger views than others in this regard. In Sweden, for example, all brand names for pharmaceutical products have to be approved by the Welfare Board before they can be used; this obviously takes care of unregistered as well as registered trade marks. The Board is concerned to see that no trade mark is used in the pharmaceutical field that could lead in its opinion, to any serious confusion with either other branded products or existing generic names. A somewhat similar situation occurs in the United Kingdom where the Medicines Commission has considerable powers under Part V of the Medicines Act. In this case the Department of Health and Social Security (DHSS) as licensing authority examines all trade marks proposed to be used on products for which Medicines licences are requested and expresses its view thereon to the Medicines Commission. It takes account of all names, registered or unregistered, and considers issues such as confusability from the safety aspect as well as general suitability. It is argued, in some circles, that this is a job more properly the province of the Trade Marks Registry. An alternative view is that while the Trade Marks Registry does an excellent job in the United Kingdom it does not have any control over unregistered trade marks. Furthermore, the DHSS is in possession of much more detailed information on the nature of the proposed product than could possibly be either communicated to or, indeed, appreciated by the Trade Mark Registration Authorities who are, after all, experts in quite a different field. The right to register and claim ownership of a trade mark is not, by any means, an undisputed right to use the mark – at its best it is a right to prevent others using it, or something similar to it, on like goods.

Trade marks are subject to all kinds of different laws and examination procedures or lack of examination procedures. In general terms, however, it is a reasonable statement of general law that a trade mark should not be misdescriptive of the product it serves to distinguish. In the United Kingdom this is embodied both in the Trade Marks Act and the Trade Descriptions Act and, indeed, in the Medicines Act – the relevant sections being respectively, Section 11 of the Trade Marks Act, which makes it unlawful to register as a trade mark or part of a trade mark any matter the use of which would, by reason of its being likely to deceive or cause confusion, or otherwise, be disentitled to protection in a Court of Justice; Section 34 of the Trade Descriptions Act which decrees that the fact that a trade description is a trade mark, or part of a trade mark, does not prevent it from being a false trade description (subject to certain provisos); and Section 85 (sub-section 5) of the Medicines Act which lays down that no person shall sell or supply a medicinal product which is labelled or marked so as to falsely describe the product or is likely to mislead as to its nature or quality or as to the uses or effects of medicinal products of that description. The law in other countries is not always so clearly stated but, as a general proposition, I would suggest the general judicial consensus of opinion is of the same ilk in most countries. Within the European Community (EC), for example, Directive 65/65 and the related complex of Directives broadly classified as 'The Rules Governing Medicaments in the European Community' might be cited as examples of this type of control, whether specifically expressed or merely implied. For example, a proposed Amendment to Article 5 of 65/65 reads '... marketing authorization shall be refused if the name of the proprietary product risks causing confusion with a proprietary product which has already been registered and which has a different qualitative composition as far as the active principles are concerned or if the name of the proprietary product is liable to be misleading as to its qualities or properties ...'.

Thus, although it would be relatively easy to formulate constructive criticism of the system by reference to actual examples of confusion and misuse, etc., it would seem that in this imperfect world, trade marks seem to be subject, at least in theory, to a variety of restraints designed to remedy most abuses – even though the system may not work in practice as well as one might wish in theory.

Generic names, quite unlike trade marks, are conceptually *not* private property. They are available for anybody to use provided they use them properly. They are orphans in a big, unfriendly, predatory world. This is their great weakness: whose responsibility are they? Because certain regulatory and similar bodies devise, propose and recommend them many people seem to believe that these bodies must own them and have some effective control over them but in this they are mistaken. A sadder and more harassed body of people than are charged with the job of devising generic names it would be hard to find in regulatory circles. One job they do on the whole well, the principal job they are supposed to do: to devise the names. This they do with enormous expertise, obeying self-imposed rules formulated to assist the scientific world for whom these names are primarily intended, listening to the wishes of the patentee manufacturers and endeavouring to steer their flimsy barque between the Scylla and Charybdis of conflicting trade mark rights on one side and scientific acceptability on the other, and a hazardous and uncertain journey it is. Their job is a formidable one: to find names that conform with rules that give them some kind of scientific sense, yet are free of trade mark conflict and not too similar to other common names both in theirs and related fields and yet are acceptable in all countries and languages to both scientists and traders. Few trade mark owners ever face such problems or achieve such results – for after all who cares much (apart from the EC Commission which has an obsession with the fancied problems) if a trade mark changes from one country to another? Why should anyone much care if cephalexin is called ORACEF in Germany and KEFLEX in the United Kingdom so long as the generic name is on the label. But if a generic name changes the result can be confusion in the literature and possibly fatality in the ward or disaster in agriculture. How then can such names be effectively devised and protected?

The principal bodies concerned in this particular field are the World Health Organization (WHO) (for pharmaceuticals) and the International Organization for Standardization (ISO) (for agricultural products such as pesticides and the like). The methods which they employ to devise their names differ. Three major problems each with some subdivisions, still remain after these organisations have devised suitable names. These are:

1. The resolution of potential conflict with:
   (a) other already devised or used common names, and
   (b) privately owned trade marks, in use or protected by law not only in the manufacturer's country, but possibly in many other countries.
2. Acceptability to the subscribing member states of WHO or ISO as the case may be.
3. Protection of the name so established and accepted:
   (a) during the stage when it is still only a proposal, and
   (b) after its formal adoption as a recommended name.

Conflict with other already devised or used common names is a matter for internal resolution between the bodies concerned. Since establishing committees always desire to establish some kind of connection between names for similar substances, although the techniques for this differ, it is not unusual for common names to become increasingly similar to one another. A much greater degree of co-operation is needed between ISO and WHO than at present exists and it is suggested that much greater use could be made of computer techniques if funds were available. Common names are, on the whole, utilised by scientific people rather than commercial people and their customers (who tend to use trade marks). For this reason the similarities are perhaps more acceptable since confusion is less likely in such circumstances. However, in the view of this author, were common names gradually to take over from trade marks, as some experts have proposed, the situation would need much more careful policing.

Conflict with trade marks is a much more serious problem. Trade mark owners are far less likely to agree to co-existence with a similar common name. There can be nothing more disastrous to a trade mark owner than having his name diluted in such a way if the goods concerned are similar; in one way or another all pharmaceutical products could be considered 'similar'. The devising bodies are well aware of this and have strenuously endeavoured to build up good will by meticulously respecting the rights of prior owners. They carefully review their proposals against specially commissioned searches of selected registers in principal countries and contact possible objectors of their own volition. As an additional precaution, proposals to adopt such names are now widely advertised in suitable journals in order that objections may be received in adequate time. Additionally, the devising bodies do their own researches as best they may (and they rely almost entirely on voluntary effort as they are almost totally lacking in funds) as to the real validity of potential objections. They examine whether the names concerned are really in use, for how long they have been apparently unused, the nature of the goods to which they are applied and the manner in which the goods concerned are used. If they feel there is a real possibility of danger they will always err on the side of caution for they have no desire to create a dangerous situation and no desire to find all their work brought to nothing because a really genuine objection makes it essential to change the name into which they have put all their hard and voluntary work. Having cleared this hurdle successfully, the names are then submitted to the Member States with requests couched in various ways to give them whatever protection may be available in the various national systems, the principal object being to prevent the names being registered as trade marks by private organisations. On the whole, names which have thus acquired recommended status have been successfully protected, although some countries have permitted some very close approximations to the officially recognised generic name to acquire trade mark registration. Registrations of this type are not really a great deal of use since conceivably they cannot prevent the use of the genuine common name side-by-side with the similarly named branded product. The principal theoretical danger period occurs in the hiatus between proposal and adoption when the name is left shivering and unprotected. If somebody stepped in and registered at this stage it is difficult to know just what preventive steps could, in fact, be taken. Possibly some kind of charge of bad faith could be brought, but one of the problems facing the inventing bodies is that they are so seriously inhibited by lack of funds. This particular danger, however, apparently has not occurred in any very serious form so perhaps this world of ours is inhabited by less evilly inclined, or perhaps less wide-awake, people than sometimes is imagined.

In Britain and, indeed, most other states, once these names have been granted recommended status and notified to the appropriate governmental departments they are effectively deemed to have passed into the language and no longer become any more acceptable for private registration for goods of like description than would any other generic word such as 'bread' or 'furniture'. If an applicant requests registration of a trade mark which, in the view of the Office at any rate, so closely resembles such a recommended name as to be obnoxious, then the British Act enables the Registrar to refuse registration on various grounds, although, of course, the applicant would always have the right of appeal. In that event his case would be decided in the light of all the circumstances of the particular case at issue.

The question of the degree of protection to be expected by common name creators for the specific syllables they have carefully inserted into the names in order to act as signposts to scientists as to the nature of the substances concerned still remains. The United Kingdom Trade Marks Registrar has agreed to take note of the various syllables selected for this purpose by the British Pharmacopoeia's Nomenclature Committee (these, basically, are the same as those set aside for the same purpose by WHO and are, therefore international in character). A list of these syllables is published by WHO under the title *General Principles for Guidance in Devising International Non-Proprietary Names of Pharmaceutical Substances*. When applications for pharmaceutical goods containing such syllables are received at the United Kingdom Trade Marks Registry the Registrar will almost certainly insist that any protection given to the proposed trade mark shall be limited to goods of the type which the syllable in question is intended to designate. As an example, if the trade mark incorporates 'bol' the goods would be limited to anabolic steroids, or 'cef' to antibiotics being derivative of cephalosporanic acid. The phrase 'almost certainly' has been used because circumstances alter cases, as always; but this is the general principle. The Registrar has a duty to avoid the registration of inherently deceptive trade marks.

Outside the United Kingdom, the position is somewhat unsatisfactory. When ISO finally recommend such a name to their subscribing members (virtually worldwide) they are supposed to prevent its acquisition as a trade mark once they have indicated national acceptance. In the case of WHO, which deals with pharmaceutical matters, recommendations only become absolutely binding on Member States (virtually worldwide) when they have agreed that they will be so bound by such recommendations. According to the practice of WHO, Member States which accept a request from the WHO Director General to recognise a particular name notified to them by him without any objection or qualification are considered so bound. The WHO authority rests upon WHO Constitution Article 2(a) which reads 'to develop, establish and promote international standards with respect for food, biological and pharmaceutical and similar products' and Article 23 which gives the Health Assembly authority to make recommendations with regard to any matter within the competence of WHO. International Names (known as INNs) fall within the ambience of recommendations in this respect.

CHAPTER 14

Available systems of structure representation

Dr W. A. Warr
ICI Pharmaceuticals Division, Alderley Park, United Kingdom

The main object of this paper is to describe ways of storing, manipulating and retrieving chemical structures using computers. It is expedient to consider first what use can be made of the computer in chemical research.

The scientific user is not satisfied by the simple retrieval of a single compound. He is also likely to be interested in groups of compounds with a given substructure. For example he may have found that compound X is active as an anthelmintic and he may wish to locate a group of similar compounds and link their substructures with other data. Scientists commonly wish to link structure with physical properties, biological activities, toxicological data or bibliographies. Not only will the scientist wish to consult his company's internal files but he will also wish to study information in patents, in the literature in general, in catalogues of commercially available compounds and in other data bases supplied by sources outside his own company. In some cases he may wish to search data bases of 250,000 or even five million structures.

Computerised current-awareness services are now commonplace – the chemist is alerted monthly to recent publications in his own specific field of interest.

Most chemists maintain personal files which they manually update but some would like to keep their personal files on a computer.

A scientist does not only wish to retrieve chemical and related data. He also needs systems to analyse and evaluate that data. Sophisticated computer tools are now available to aid in the interpretation of spectral data,[1,2] to perform theoretical calculations and molecular modelling,[3-5] to investigate structure/activity relationships,[6-9] to index reactions[10-16] and to aid in the design of organic synthesis.[17-19]

At the base of all the systems mentioned so far is a need to reduce a three-dimensional structural diagram to a 'name' that can be stored and manipulated by a computer. Systematic nomenclature in its usual sense is not useful here. There are many reasons why this is so. Systematic names are often long and complicated. When a new area of chemistry is opened up the appropriate nomenclature is not available for months or perhaps years and confusion reigns in the interim. The rules[20,21] for constructing names are often inconsistent or ill-defined. No-one has yet been able to write a computer program which will write systematic names for a wide range of given structures because systematic nomenclature rules are not simple, logical and rigorous. In the International Union of Pure and Applied Chemistry (IUPAC) nomenclature there is inconsistency in the retention of trivial names and in some cases there may be more than one acceptable name for a single compound. In a useful computerised system any one compound must have only one name and any name must be convertible to only one structure. Chemical Abstracts Service (CAS) have systematised nomenclature sufficiently to give a unique name for most compounds[21,22] but in doing so they have had to make even more rules (and change their own rules) and there are now two acceptable forms of nomenclature to recognise, CAS and IUPAC.

While systematic nomenclature has disadvantages as a unique identification tool for the storage of structures it has even greater disadvantages when substructure searching is considered. All the structures in Figure 1 contain N—C=N. It should be obvious what an enormous range of names would be needed as search terms in an enquiry for compounds containing N—C=N.

Since nomenclature has these disadvantages, other methods of structure representation have been examined. Table 1 shows some of these methods. The

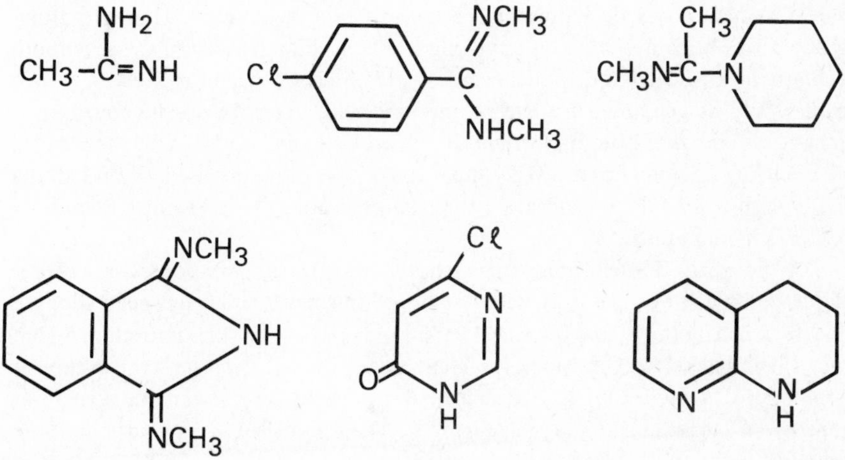

Figure 1

scientist's preferred tool of communication is the two- or three-dimensional structure diagram. He will frequently use molecular formula as a literature search tool, even though it does not completely represent structural topology, because

Table 1 – Chemical structure representation

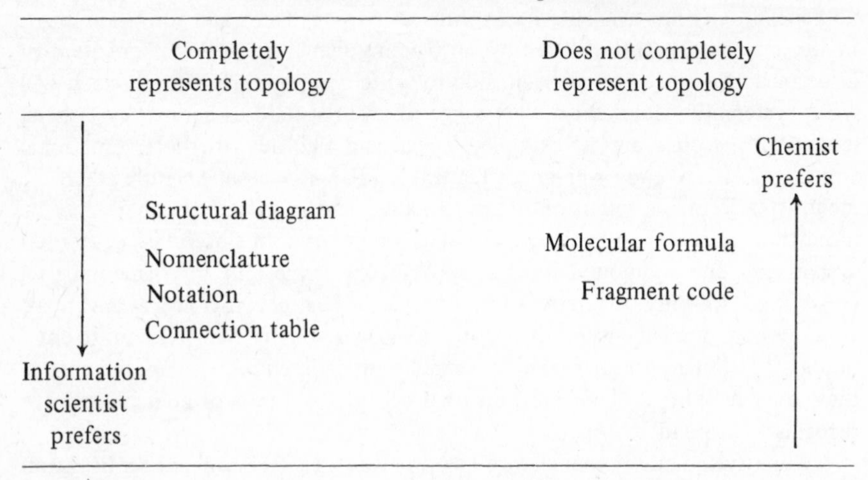

Completely represents topology	Does not completely represent topology
	Chemist prefers ↑
Structural diagram	
Nomenclature	Molecular formula
Notation	Fragment code
Connection table	
↓ Information scientist prefers	

he may feel that systematic nomenclature is becoming too complex for the average scientist. The information scientist, on the other hand, prefers to use fragment codes, notations and connection tables in a mechanised system.

A fragment code is a collection of predetermined small substructures (e.g. six-membered ring, carbonyl group) each represented by a number, or sometimes a combination of letters and numbers. In the early days of fragment codes each fragment could be represented by a hole in a punch card. Thus a hydroxy group might be represented by punching a hole in the third row of the seventieth column of an 80-column punch card, represented on paper as 70/3. A punch card would be generated for each compound with holes for all the appropriate fragments and the structure would be drawn by hand on the card. Mechanical card sorting equipment (or even simple rods) could then be used to find all the cards with a hole 70/3 (and any other required holes) to perform a search for hydroxy-compounds.

A group of German companies, the Pharma Documentation Ring (PDR), developed a code of this type with about 300 fragments and it has been adopted and modified by the Derwent Central Patent Index for use in their Farmdoc, Agdoc and Chemdoc services[23]. Since fragment codes do not represent the complete topology of a molecule they are particularly useful for the variable structures that often occur in patents, e.g. Figure 2. These so-called Ring Codes are now searched more efficiently by computer.

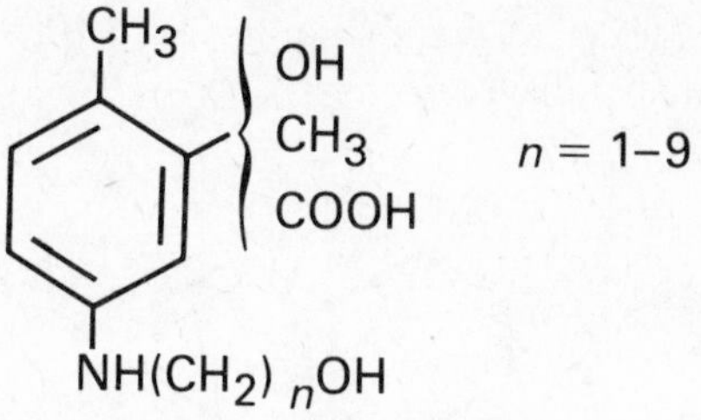

$n = 1\text{–}9$

Figure 2

The Pharma Documentation Ring has extended Ringcode for use in reaction indexing.[10,24] A further system called GREMAS (generic retrieval by magnetic tape storage) has been developed by International Documentation in Chemistry using a more complicated fragment code for reaction indexing.[24]

The Duffin fragment code[25] and the ICI CROSSBOW (computerised retrieval of organic structures based on Wiswesser)[26-28] fragment code are other well-known examples. In the latter, 152 fragments can be stored as on/off switches on the computer in the space of only nineteen characters and it is quick and efficient to search the codes (200,000 are searched in less than a minute).

In modern systems the fragment codes for a compound are usually generated by the computer from a notation or connection table.

Most of the common fragment codes are predefined. For structure–activity studies an open-ended fragment code is more useful but details of such work are beyond the scope of this paper.

Fragment codes describe the various parts of a molecule but not how those parts are linked together. Therefore a substructure search of fragment codes will nearly always output a substantial number of compounds which have the required code pattern but do not resemble the required structure.

In the CROSSBOW system some or all of this 'noise' can be eliminated by searching Wiswesser Line Notations (WLNs).

WLN[29] is a compact way of uniquely and unambiguously representing the complete topology of a molecule by a linear sequence of characters. An example is given in Figure 3. Several notations have been invented over the years but only

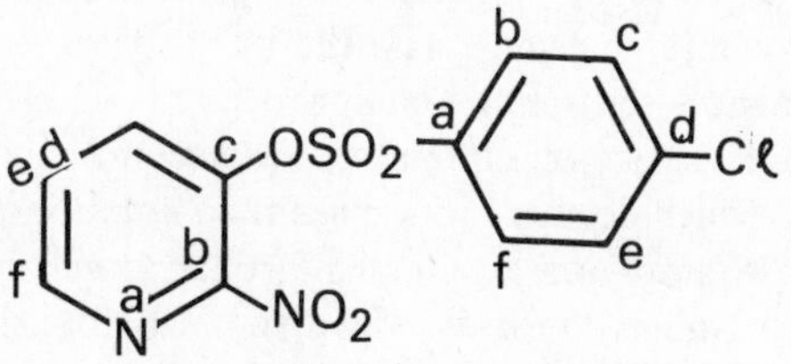

Pyridine = T6NJ
Phenyl = R
Chlorine = G
Dioxo = W

T6NJ BNW COSWR DG

Figure 3

WLN has gained world-wide acceptance. For interest a WLN and a Dyson/IUPAC notation[30] are compared in Figure 4.

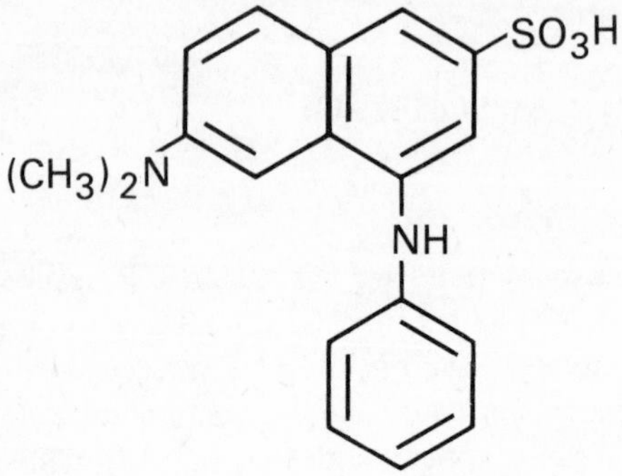

SO_3H

$(CH_3)_2N$

NH

IUPAC: $B6_2N(C)_2 4SO_2Q8$:ION/B6

WLN: L66J BMR& DSWQ IN1&1

Figure 4

An acceptable notation system must have the following characteristics. The notations should be amenable to automatic storage and retrieval. The encoding rules should be simple, logical and well-defined. All, or almost all, classes of compound should be covered. The notation should be as compact as possible. Any one notation should be decodable to only one structure and there should be a unique notation for any one structure. It also helps if the notation is made more readable by the occurrence of words and spaces and if chemically significant fragments can be recognised.

To a large extent WLN meets these criteria. However, it has quite a number of the disadvantages that systematic nomenclature has and it does not handle tautomerism, mesomerism or stereochemistry well. In common with all notations and systems of nomenclature, it is not 'user-friendly'. However, it is a succinct way of recording a structure and WLN-based systems have proved cheap and efficient over the years in a wide range of organisations. WLNs are readable, listable, permutable and useful for interchange of data the world over.

In the early days of information science the compactness of WLN was a great advantage. However, as storage space on computers has become relatively less expensive, connection tables, which occupy more space than notations, have come to the fore. Table 2 is an example of a redundant connection table. It is termed redundant because each connection is described twice. Before storing such a table the computer would compact it (i.e. remove the redundancy) and use some logic known as a Morgan algorithm[31] to make a unique connection table for the given structure. CAS use unique, bond-explicit, connection tables

Table 2

$$\begin{array}{c} \overset{4}{O} \\ \| \\ \underset{1}{CH_3}-\underset{2}{CH_2}-\underset{3}{C}-\underset{5}{NH_2} \end{array}$$

Atom number	Atomic symbol	Bond connection	Attached atom number	Bond connection	Attached atom	Bond connection	Attached atom
1	C	1	2				
2	C	1	1	1	3		
3	C	1	2	2	4	1	5
4	O	2	3				
5	N	1	3				

of this type. Clerical labour can be used to input structural data in such systems. The CROSSBOW system is rather different. Skilled labour is used to input WLNs and the CROSSBOW connection table is generated from the WLN by the computer. The table is automatically unique because the originating WLN was unique. A CROSSBOW connection table is bond-implicit. Its basic unit is an atom-plus-its-bonds. Its atom and bond information requires less storage space than in a bond-explicit connection table but the information is harder to unscramble. However, the CROSSBOW connection table does have extra useful information such as details of rings and their type.

One advantage of using connection tables is that they allow the possibility of atom-by-atom search. A network such as the —N—C=N—, quoted earlier, can be matched against every connection table on a file to find those compounds which contain the required network. In practice, to search every compound in a large file in this manner would be over-expensive. Usually a fragment code search or a string search of WLNs or molecular formulae will be carried out first to reduce the file to manageable proportions.

Another great advantage of connection tables is that they can be computer-derived from structure input and they can be used to output computer-drawn chemical structures. The chemist's preference for a structure in/structure out system can be met by a system based on bond-explicit connection tables. Moreover, stereochemically extended Morgan algorithms have been written for systems such as Molecular Access System (MACCS)[32] and Simulation and Evaluation of Chemical Synthesis (SECS),[18,19] and three-dimensional structure input and output is now possible.

Structures are input to MACCS using a light pen and a graphics terminal. The user is able to input his request almost as if he were drawing a structure on paper. He can use stereochemical wedged and hatched bonds and he can call on a 'menu' of predrawn structures to reduce the effort of input. MACCS can then perform a substructure search. Input to SECS is similar but the system predicts possible reaction precursors and allows the user to interact and demand forerunners for his preferred precursors.

MACCS and SECS have been chosen as examples. Also available are Logic and Heuristics Applied to Synthetic Analysis (LHASA)[17,19] to help with organic synthesis, and Description, Acquisition, Retrieval and Correlation (DARC)[33] to search 4.5 million Chemical Abstracts compounds while-you-wait and the National Institute of Health's Chemical Information System[34,35] which can access an assortment of files (e.g. crystal structure and mass spectral data) and rotate accurately modelled output structures.

CAS now has its own on-line system.[36] CAS is noteworthy for having sophisticated ways of handling enormous quantities of data, for having text-handling programmes which permit the use of systematic nomenclature and for having Registry Numbers which are becoming accepted as an alternative to names.

All these systems are connection table based. The systems of the future are likely to make best use of the scientist's preference for using three-dimensional structural diagrams and the information scientist's preference for connection tables (and probably Chemical Abstracts Registry Numbers) while increasing the amount and variety of information that can be searched and evaluated.

REFERENCES

1. R. E. Carhart, D. H. Smith, H. Brown and C. Djerassi, *J. Am. Chem. Soc.*, 1975, **97**, 5755.
2. *Computer-Assisted Structure Elucidation*, ACS Symposium Series Number 54 (ed. D. H. Smith).
3. R. Langridge, T. E. Ferrin, I. D. Kuntz and M. L. Connolly, *Science*, 1981, **211**, 661.
4. P. Grund, J. D. Andose, J. B. Rhodes and G. M. Smith, *Science*, 1980, **208**, 1425.
5. G. R. Marshall, H. E. Bosshard and R. A. Ellis, Computer modelling of chemical structures: Applications in crystallography, conformational analysis and drug design, *Computer Representation and Manipulation of Chemical Information* (eds. W. T. Wipke, S. R. Heller, R. J. Feldmann and E. Hyde). John Wiley, 1974.
6. C. Hansch, A. Leo and D. Elkins, *J. Chem. Doc.*, 1974, **14**, 57.
7. A. Leo, D. Elkins and C. Hansch, *J. Chem. Doc.*, 1974, **14**, 61.
8. D. Elkins, A. Leo and C. Hansch, *J. Chem. Doc.*, 1974, **14**, 65.
9. S. M. Free and J. W. Wilson, *J. Med. Chem.*, 1964, **7**, 395.
10. *Chemical Reactions Documentation Service* (now on-line). Derwent Publications, Rochdale House, 128 Theobalds Road, London WC1.

11. M. Osinga and A. A.V. Stuart, *J. Chem. Doc.*, 1974, **14**, 194.
12. M. Osinga and A. A.V. Stuart, *J. Chem. Inf. Comput. Sci.*, 1976, **16**, 165.
13. M. Osinga and A. A.V. Stuart, *J. Chem. Inf. Comput. Sci.*, 1978, **18**, 26
14. M. F. Lynch and P. Willett, *J. Chem. Inf. Comput. Sci.*, 1978, **18**, 149.
15. M. F. Lynch and P. Willett, *J. Chem. Inf. Comput. Sci.*, 1978, **18**, 154.
16. M. F. Lynch, P. R. Nunn and J. Radcliffe, *J. Chem. Inf. Comput. Sci.*, 1978, **18**, 94.
17. E. J. Cory and W. T. Wipke, *Science*, 1969, **166**, 179.
18. W. T. Wipke, Computer-assisted three dimensional synthetic analysis, *Computer Representation and Manipulation of Chemical Information* (eds. W. T. Wipke, S. R. Heller, R. J. Feldmann and E. Hyde). John Wiley, 1974.
19. *Computer-Assisted Organic Synthesis*, ACS Symposium Series Number 61, (eds. W. T. Wipke and W. J. Howe).
20. *Nomenclature of Organic Chemistry*, Sections A, B, C, D, E, F and H: 1979 edition. Pergamon Press, 1979.
21. *Chemical Abstracts Index Guide, 1977,* Appendix IV paras. 101–306.
22. *Parent Compound Handbook,* American Chemical Society (obtainable from Chemical Abstracts Service, PO Box 3012, Columbus, Ohio 43210, USA, or from United Kingdom Chemical Information Service in the UK).
23. *Ringdoc Instruction Bulletins* Nos. 5, 6 and 7, Derwent Publications, 1964-1972.
24. J. Valls and O. Schier, Chemcial Reaction Indexing, *Chemical Information Systems* (eds. J. E. Ash and E. Hyde). Ellis Horwood, 1975.
25. W. M. Duffin, *J. Chem. Doc.*, 1961, **1**, 44.
26. D. R. Eakin, The ICI CROSSBOW System, *Chemical Information Systems* (eds. J. E. Ash and E. Hyde). Ellis Horwood, 1975.
27. D. R. Eakin, E. Hyde and G. Palmer, *Pesticide Sci.*, 1974, **5**, 319.
28. E. E. Townsley and W. A. Warr, 'Chemical and Biological Data – an Integrated On-line Approach', paper given at Spring 1978 American Chemical Society Meeting published in ACS Symposium Series Number 84.
29. E. G. Smith and P. A. Baker, *The Wiswesser Line-Formula Chemical Notation* (*WLN*), 3rd edn, Chemical Information Management Inc., New Jersey, USA.
30. G. M. Dyson, The Dyson–IUPAC Notation, *Chemical Information Systems* (eds. J. E. Ash and E. Hyde). Ellis Horwood, 1975.
31. H. L. Morgan, *J. Chem. Doc.*, 1965, **5**, 107.
32. *The Molecular Access System of Molecular Design Inc.*, Haywood, California, USA.
33. Information on DARC can be obtained from R. Attias, Centre National de l'information Chimique, 25 Rue Boyer 75971, Paris Cedex 20, France.
34. S. R. Heller, G.W. A. Milne and R. J. Feldmann, *Science*, 1977, **195**, 253.
35. G. W. A. Milne, S. R. Heller, A. E. Fein, R. G. Marquart, J. A. McGill, J. A, Miller and D. S. Spiers, *J. Chem. Inf. Comput. Sci.*, 1978, **18**, 181.
36. N. A. Farmer and M. P. O'Hara, *Database,* 1980, 10.

CHAPTER 15

Sources of help with chemical nomenclature

Dr D. S. MAGRILL
Beecham Pharmaceuticals, Brockham Park, United Kingdom

There are particular problems in using chemical nomenclature which affect chemists and information scientists who lack highly developed chemical nomenclature skills. Names may be assigned to chemical compounds for a variety of reasons. It may be necessary to refer to a compound during a discussion or in a written document; it may be necessary to describe a compound's structure by means of its name; it may be necessary to decide where in an index to look for a reference to a compound; or it may be necessary to give a unique and unambiguous name to a compound so that it can be confidently assigned to its place in an alphanumeric index. These various requirements impose different standards of rigour on the user and no-one should waste intellectual effort assigning to a compound a more 'correct' name than the circumstances warrant – particularly if there is a risk that his best efforts might still yield an incorrect result.

For example, no chemist wishing to describe his work informally to his colleagues would consider assigning International Union of Pure and Applied Chemistry (IUPAC) names to the compounds he has prepared. He would not even use these names if they were written on the bottles. Instead he would use a pragmatic assortment of trivial names, semi-systematic names and structural diagrams and he would derive a two-fold benefit. He would be able to name compounds with ease; and his colleagues would understand him.

Although this may be self-evident to the point of triviality in the context of the laboratory or coffee-room discussion, it often gets overlooked in more formal scientific intercourse. A written report for circulation within a scientist's organisation, for example, is still designed to communicate information and common names or structural diagrams are the way to achieve that. A paper in the scientific literature should fall into the same category. The descriptions assigned to chemicals here fulfil two functions – communication and archiving – but there is no need for an archive to resort to tongue-twisting, brain-bending nomenclature. If a description of a compound is fully understandable by the reader then it should usually serve also as a wholly adequate archival record even if it is no more than a phrase like 'the bromoketal, IV' accompanied by a structural diagram. Names of this sort have long been employed by chemists for the best of reasons: they are easy to use; they are easy to understand; they give an additional indication of the features of the structure which are significant in the context; and, as a bonus, they even provide an element of self-checking.

If editors of prestigious journals ask chemists to use IUPAC or Chemical Abstracts Service (CAS) nomenclature it should be resisted unless it adds to the clarity or precision of the paper – a somewhat improbable occurrence.

The same reasoning applies to chemical patents. Here again the prime objective is to identify without ambiguity the entity under discussion and again a structural diagram is usually the safest recourse. A company which preferred to rely on CAS or IUPAC names might almost be accused of trying to hide the nature of its inventions, so obscure are many of the names produced by these systems.

There are occasions when a chemist really does need to assign names. Such occasions fall into two broad groups. First, it may be that for some legal, regulatory or other purpose a chemist will require a name which is, at least, valid and unambiguous and the advantage of adhering to IUPAC or CAS convention in these circumstances is that a name so derived is very unlikely to contain unrecognised ambiguity. The other broad circumstance is perhaps more common. It arises when a scientist needs to know where to look for a compound in a list or index of some sort, perhaps in a list of toxic substances, a list of compounds subject to certain legislation or a major index such as the *Chemical Abstracts* chemical substance index.

Where can the user turn for help? The simplest form of help is direct assistance from an expert. In the United Kingdom there are two groups of experts who can be contacted for help. The Laboratory of the Government Chemist operates a Chemical Nomenclature Advisory Service with staff who are particularly skilled in IUPAC nomenclature. Assistance is provided over the telephone for simple enquiries or in writing for more complex queries. A modest charge is applied where appropriate. This service is useful for those who need a valid, unambiguous name for a compound.

Help with CAS nomenclature is obtainable from the Royal Society of Chemistry's Chemical Information Service (formerly UKCIS) at Nottingham. As national *Chemical Abstracts* agents, abstractors and indexers, this group can provide telephone or written advice on CAS nomenclature and chemical information in general. Such direct methods cannot provide the answer to all problems. Advisory groups would soon become swamped if users all telephoned or wrote to them every time there was a doubt about what to call a compound, or where to look for a material in *Chemical Abstracts*. Moreover, confidentiality considerations may preclude the use of such sources.

For known compounds there are a number of printed lists one might consult. A large collection of compounds important enough to have trivial, approved or semi-systematic names are listed in the *Index Guide to Chemical Abstracts* along with their CAS name. Table 1 gives a number of other lists that may be useful.

Table 1 – Useful directories of compound names

Index Guide to Chemical Abstracts
Merck Index
Handbook of Chemistry and Physics (Rubber Handbook)
Register of Toxic Effects of Chemical Substances (NIOSH)
USAN and the *USP Directory of Drug Names*
Eurorepertoire of Chemicals

The next stage would perhaps be to use the most common means of locating a compound name in *Chemical Abstracts*, a search of the molecular formula index. It has to be assumed when using this index that even if a chemist could not write the name of the compound he would, at least, recognise it if it could be seen. Most users can manage this – or at least manage to eliminate most of the irrelevant names – but it may be quite difficult at times. Figure 1 shows the CAS name and structure of a comparatively simple compound. The name is derived by routine application of logical rules and is not particularly complex. Even so, a chemist looking for references to the compound in the molecular formula index of *Chemical Abstracts* might be greatly exercised to convince himself that this name, with its five levels of parenthesis, really refers to the compound he wants.

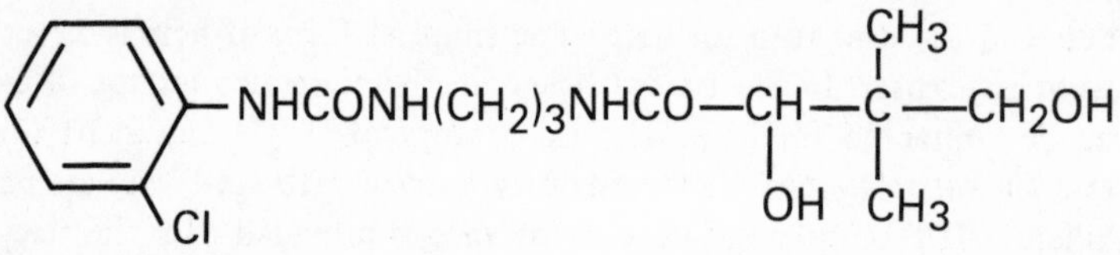

CH_3

$NHCONH(CH_2)_3NHCO—CH—C—CH_2OH$

Cl OH CH_3

Butanamide,

N-[3-[[[(2-chlorophenyl)amino]carbonyl]amino]propyl]-2,4-dihydroxy-3,3-dimethyl

Figure 1

While on the subject of hard-copy sources of help, it is useful to consider the traditional method of last resort – reading the instructions. In most contexts it may be facetious to regard this as a method of last resort but not in the field of chemical nomenclature. The IUPAC rules and all derived, related and subsidiary rules, are scattered amongst various papers and publications. To collect them is a daunting task; to read them is far worse. A summary in the *Chemical Abstracts Index Guide* covers all the rules, albeit very sketchily, and gives a complete bibliography. Of the books on the subject, Cahn and Dermer's *Introduction to Chemical Nomenclature*[1] goes a long way towards making sense of a difficult topic but it would be unrealistic to expect it to do more than ease the reader into the subject.

One useful source of chemical information which does not rely at all on orthodox nomenclature is the Index Chemicus Registry System (ICRS). This is a data base of core organic chemistry literature where all novel compounds are indexed in a number of ways including Wiswesser Line Notation (WLN). The location of a particular compound in ICRS system is through the WLN and this is usually an easier process than with other systems. The notation can be sought directly in a permuted WLN index where other related compounds are listed. More sophisticated substructure searching is available through the use of computer tapes although these are not widely accessible because of their high cost. If the user still needs to know the correct name of the compound that is being sought – or of a similar compound from which it may be possible to infer the name of the unknown – ICRS would be of little direct help. It would give the user a reference to an original publication that might name the compound properly but in all probability it would be necessary to retrieve the corresponding *Chemical Abstracts* reference in an online CA Search file (using the bibliographic data found from ICRS) and to print the Registry Numbers. The corresponding systematic names could then be ascertained either from dictionary files or from the printed *CAS Registry Number Index*. The overall procedure is oulined in Figure 2.

The most familiar online dictionary files are Chemline, Chemname and Chemdex, accessible on BLAISE, Lockheed and SDC respectively, which are intended as an adjunct to bibliographic searching. The Lockheed files are the most complete, containing between them every compound mentioned in *Chemical Abstracts* since 1972 – a total of 3.5 million compounds. The user can approach a compound in a Lockheed dictionary file in a number of ways. It can be considered by synonym, trivial name or approved name; it can be

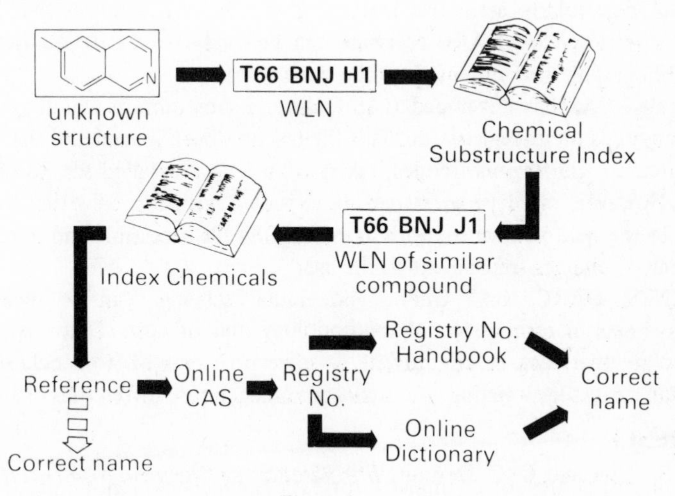

Figure 2

searched by molecular formula and combined with, perhaps, a fragment or two that the chemist is confident will occur somewhere in the correct name. Often the user can specify details about the number, size, composition and groupings of rings. It is relatively simple using these keys to find a particular compound, and hence its name and Registry Number, in a Chemname file. Acyclic analogues present a much greater problem than the cyclic analogues.

There are a number of systems which can, in effect, identify compounds from simple structural descriptions. The earliest in the field was the Structure and Nomenclature Search System (SANSS) used by the US Government sponsored databank collection, CIS. This system enables the user to draw a structure or a substructure by using a set of simple commands on a teletype terminal and then to search all the collections for compounds that match the query structure. The search procedure is a little ponderous but it provides a simple and reliable means of finding a compound without making any effort to name it and the system will display the structures of all compounds it retrieves along with their Registry Numbers and current CAS names.

A serious limitation with SANSS is that its compound coverage is restricted to about sixty collections comprising only a few hundred thousand compounds. The Description, Acquisition, Retrieval and Correlation (DARC) chemical retrieval system available online via Euronet from Telesystemes-Questel has much more comprehensive coverage and is easier to use. DARC is essentially an online dictionary file covering all structurally defined compounds on the CAS Registry file – over four million of them. It is searchable by structure and substructure. Query structures are easily described in simple topological terms and the output can be displayed in clear graphical format to users with a suitable graphics terminal. Users with a simple teletype terminal are limited to Registry Numbers. In most cases a laborious hard copy operation is then needed to retrieve the names. A particularly attractive feature of the Questel system is that Registry Numbers retrieved by DARC software can be used directly for queries on the CAS bibliographic files mounted on the same system.

Finally, CAS has developed CAS Online, a substructure search system for the compounds on its registry file. This file has now been loaded and the searcher is required to use fragment codes, taken from a very detailed set, to effect his search. However, rapid progress can be expected with this substructure search system in the next year or so, both in making the file accessible and in rendering the search mechanics transparent to the user.

SANSS, DARC, CAS Online and similar schemes can be regarded as major systems in terms both of methodology and of cost. That users should contemplate their use to circumvent or solve problems of nomenclature is an indication of just how vexing and intricate nomenclature difficulties can be.

REFERENCE

1. R. S. Cahn and O. C. Dermer, *Introduction to Chemical Nomenclature,* 5th Edition, Butterworths, 1979.

CHAPTER 16

Harmonisation of nomenclature systems

Dr S. E. WARD
Glaxo Group Research Ltd, Ware, United Kingdom

Other contributors to this book have indicated the range and complexity of problems of chemical nomenclature. The current position could perhaps be summarised with the statement: 'Chemical nomenclature and the naming of chemicals is still in chaos'. Nevertheless, users still have to face the problem that the same chemical substance will acquire a variety of names during its lifetime depending on its popularity and, more irritatingly, on the idiosyncrasy of the scientists who synthesise, investigate, and use it. In addition it is necessary to try and scheme a way forward which will simplify the future and avoid perpetuating the present complexity.

The following nomenclature systems have been identified.

Systematic nomenclature which is based on the combination of a series of terms, each of which has a definite and constant meaning in terms of chemical structure, to produce an unambiguous (not necessarily unique) nomenclature. There is still, however, no single set of universally acceptable rules and the devotees of Chemical Abstracts (CA) and the International Union of Pure and Applied Chemistry (IUPAC) continue to cause confusion in the primary literature and in indexes to the primary literature. In the case of CA, the changes in nomenclature, albeit to more rather than less systematic conventions, have been dominated by the motive of efficient publication rather more than by the needs of the average user.

Semi-systematic nomenclature which often serves in verbal communication and as a shorthand for written communication, and preserves some relationship between structural analogues, illustrating features which seem important for the particular purpose at the time. It derives partly from systematic and partly from trivial nomenclature and this in turn originates partly from history, partly from popularity, and is used by scientists and non-scientists alike.

Trivial nomenclature which includes circumstantially assigned names, e.g. source names, and also includes trade names or proprietary names which are subject to some legal monitoring and may identify a single compound or a mixture. The term is often used to cover non-proprietary or generic names for compounds exhibiting certain functions, e.g. pesticides and drugs, which are controlled frequently by national legislative committees, and which, in theory, provide a method of identifying a substance likely to have several other names. Assignment of generic names is to some degree systematic since attempts may be made to retain certain name fragments across a series of structures with a common property, e.g. -azepam for substances analogous to diazepam (BAN rules).

Overlying the complexity of these three nomenclature levels is the complication of misapplication of systematic rules,[1] alternative spelling conventions (ph for f), spelling mistakes, and the fact that the rules change, leaving behind the current favoured name a collection of historical antecedents – all of which may be keys to valuable information on the compound.

The nomenclature picture is therefore sophisticated and, in addition to alphabetic and alphanumeric names, a variety of methods exist by which chemical substances can be identified and chemical structure expressed.

Identification numbers are assigned by industry for internal housekeeping, by national and international bodies for simple identification, and by publishing organisations such as CA for control and the simplification of computer processing of information on chemical structures.

Notations such as Wiswesser Line Notation (WLN) have developed to describe the two/three-dimensional structure diagrams in a form suitable for computer processing as an alphanumeric character string and are used both within industrial files to document structures synthesised and as an indexing tool for published information.

The connection table, a computer processable record describing the spatial arrangements of atoms and molecules, provides the most basic record of molecular structure and is used increasingly as a key storage record in chemical information systems, as are the direct representations of structural diagrams in digital form which provide a means of high-quality structure reproduction.

Fragment codes bear a direct relationship to nomenclature since they are based on the assignment, either manually or automatically, of symbols to regularly occurring structural fragments. Fragment codes are not, of course, a complete description of a molecule: they identify important features only and do not normally give any clue as to the relative arrangement of these in the molecular structure. They also fail to take account of the different behaviour of structural groups in different molecular contexts (a failing shared by all varieties of nomenclature).

This then is a brief overview of the current picture and certainly in the field of machine processing of chemical structures the picture is not static since new methods of representing chemical structures uniquely and unambiguously continue to be published with alarming regularity. For instance, Dr Walentowski of Beilstein has produced recently[2] a structure-nomenclature notation, SNN, which specifies a molecule by a series of fragments linked together by special signs; the description is then used to generate a code number by which the compound can be assigned its proper place within the Beilstein classification.

More and more methods are available for the description of chemical structures and the implementation of these will result in greater and greater confusion in the communication and accessibility of information on chemical structures. Even a simple compound can acquire an enormous number of names. Likewise the quantity of chemical structures to be identified grows apace – from 1,313,848 compounds featured in CA registry files in 1969 to 4,787,991 in 1979 – and continues to expand significantly.

Chemical nomenclature and its alternatives can be seen therefore as continuing to be ill-disciplined and uncontrolled – scientific and trade independence flourishes and evolves towards wider differences. How then can the situation be rationalised? Is any degree of harmony in fact possible? Harmonisation is not standardisation and it is possible for both different people and different nomenclature systems to exist side by side in harmony. It is, however, important to examine which nomenclature systems are worth retaining, which levels might be eliminated and how harmony between the remaining systems can be achieved. This can only really be done if the reasons for naming compounds are examined with a reasonable degree of objectivity. No reduction in the variety of names for a compound is possible without willpower and discipline. It is useful to consider the reasons for naming compounds, and the uses and the users of those names, in the hope of developing a rational analysis and solution to the problem.

The key problem is accurate capturing, communicating, storing and retrieving information on chemical molecules, most of which can be represented by a structural diagram. A substantial amount of information can exist on the pure compound itself. In addition the molecule may be combined into other molecules in a pesticide or drug formulation with its own distinct properties; the molecule may also have a series of close derivatives, e.g. salts, whose chemical and biological properties will often be extremely similar although physical characteristics may vary; the molecule will also be related to other chemicals – analogues, the definition of analogue depending on the particular viewpoint from which the compound is being considered.

This coarse chemical classification must now be set against the users of chemicals, information on chemicals, and hence currently of chemical nomenclature, and the variety of situations in which nomenclature will be used. The body of chemical information can be classified into two types: reaction information and property information. Reaction information can be defined as the chemical properties of molecules which can be used to derive routes – derived from generalising a series of specific applications, and the physico-chemical principles and data which govern successful synthesis. Under property information is included both the physico-chemical data which is used in confirmation of a structure and biological and physical properties which decide the utility of the substance.

The users of such information can be classified into four categories:

(1) the trained chemist/biochemist who is continually involved with the handling of chemical substances and is both creating and using information on them;
(2) scientists other than chemists who again will be using chemical substances and who will be concerned with properties other than chemical ones, e.g. the pharmacologist;
(3) the non-scientist: people involved in transport and storage of chemicals, the horticulturalist or amateur gardener, legislators and those enforcing legislation;
(4) the student: here there are particular problems peculiar to the acquisition and transfer of chemical knowledge.

People at all these levels may wish to communicate (verbally or in writing), acquire, store and retrieve information on chemical substances. Retrieval needs to be done in two ways

- novelty checking/single compound retrieval, i.e. the search for information on a particular chemical of known structure or with a particular name, e.g. a trade name;
- substructure/analogue searching, i.e. the retrieval of information on compounds which have a structural feature (or features) in common.

Where is this information? It may appear in primary scientific publications, be summarised from these into secondary abstracts services, be analysed and reviewed in reference books, text books, and dictionaries. It may feature in patents or other legislation, containers, etc.

Can any harmony at all be achieved across this enormously complicated jumble of a host of different nomenclature systems each with its own protagonists vying jealously for supremacy, with the diverse needs of the various levels of users, the different levels of communication and the different requirements for retrieval?

The first thing to state positively is that, unlike subjects where familiarity with syntax, vocabulary and grammar is a prerequisite to further study and understanding, for chemistry, as indeed for other sciences, a knowledge of terminology, and in particular nomenclature, is not crucial to an understanding of the behaviour of molecules, and is not even essential for their identification and recognition. The best and least ambiguous form of representation of a chemical compound is its structural formula. The structural formula is the preferred method of communication between chemists and is used whenever the chemist wishes to express the actual structure uniquely and unambiguously; chemists think of and discuss molecules and their behaviour largely in terms of two-dimensional diagrams (with certain conventions for detailing stereochemical information). It is noticeable that an extremely popular abstracts journal, *Current Abstracts in Chemistry,* relies almost entirely on the structural diagram for conveying information.

Systematic chemical nomenclature, on the other hand, is almost irrelevant to day-to-day communication between chemists: its rules are complex requiring enormous effort and constant use to achieve familiarity, and systematic names are frequently unpronounceable and totally unsuitable for oral communication. Few chemists know the rules well and will normally refer to an expert when required to present a systematic name, normally at the time of publication. The importance of systematic nomenclature in written communication, i.e. publication, has arisen mainly because of the higher cost and space required to replicate the structural formula in comparison with written text, and possibly because of the problems associated with indexing structures.

The further difficulty of the slowness with which rule bodies respond to new areas of chemistry so that a name is required before a rule is available has to some extent been overcome by the promptness with which CA have to devise systematic names for compounds appearing in primary publications.

For the research chemist then, systematic nomenclature is of limited use in communication; does it have more significance in the retrieval of information?

It might be expected that for retrieval of information on specific compounds an unambiguous and unique name of the type assigned by the CA 9th collective index rules would be superb. However, the lack of knowledge of the chemist and the history of changes to nomenclature ensure that the more easily derived and universally accepted molecular formula is normally the first search point, with its results refined by rapid scanning in order to recognise the structure of interest. Nor is systematic nomenclature any better in the retrieval of a group of analogues since in an alphabetically arranged index structural analogues are widely separated and in an hierarchical index such as the CA Chemical Substances Index, the definition of parent via the molecular skeleton which contains the highest ranking functional group, again results in the separation of related compounds. Even when machine searching can remove some of these limitations by permitting retrieval on a combination of name fragments, systematic nomenclature still has disadvantages since its syntax and the vocabulary which has already defined certain atomic arrangements will restrict the scope of queries which can be handled to searches which combine fragments without defining their spatial relationships. Nomenclature and fragment codes therefore have the same limitations as far as substructure searching goes, disadvantages which can be overcome via the connection table generated directly, or via WLN.

If systematic nomenclature has the disadvantages outlined, for storage, retrieval and communication of primary information between chemists it should be possible to reduce the reliance on, and even eliminate, systematic nomenclature as the primary method of recording and indexing chemical structure, replacing it with more effective services from the structural diagram itself, the connection table, the notation, and the molecular formula.

Current technology has simplified the printing of chemical structures which can therefore be the preferred form of printed communication. This already happens in many papers where nomenclature is only used to describe common reagents or to derive generic nouns for the group of compounds under consideration.

Retrospective searching (and indexing) of the primary literature by chemists (or information scientists acting on their behalf) is normally achieved using a secondary service such as CA and the Index Chemicus Registry system. Both these data bases have as their core a non-nomenclature record: in the former case a connection table, in the latter a Wiswesser line notation, both of which can absorb new areas of chemistry rapidly. Although not yet fully developed for direct on-line retrieval both data bases are evolving substructure searching facilities which are far superior to those offered by nomenclature systems and which will allow input of a structural formula (or part) to initiate an enquiry. Although neither system offers comprehensive access to the primary literature it should be possible for CA to extend its registry retrospectively to ensure its registry files include all known substances.

Nomenclature therefore reduces in importance for the practising chemist for communication and analogue retrieval. Retrieval of information on specific compounds is easily achievable via Index Chemicus and would also be achievable, although currently at some cost, through the CA structure search system. The molecular formula printed index will, however, continue to serve as a pointer to specific compounds – with nomenclature as a secondary classifier for some time.

In addition, alongside primary and abstracts literature sit the collated and assessed information in reference books, text books, etc. Here again, molecular formula should be the indexing tool of priority for retrieval of information on specific compounds. Also in text and reference books there is a need for indexing by chemical classification, i.e. generic chemical terms. Within the population using them these terms are well-known, well-controlled and not substantial in number so should remain as a semi-systematic generic nomenclature derived as is currently the case from systematic terms or automatically from connection tables.

At the educational level, the range of compounds to be dealt with is smaller. As in the case of the research chemist, much communication can be done using the structural diagram, and the retrieval problems are limited. Generic terms to describe families should be identical with those used for the research chemist. Systematic nomenclature will continue to be important in the demonstration of relationships between analogues since trivial nomenclature is a barrier to acquisition of knowledge. It is important that the system used is identical with that which will be encountered in the research and industrial workplaces.

The needs of the layman and the non-chemical scientist for chemical nomenclature in retrieval are similar. Here, of course, the chemicals of interest are small in number and will be chemicals on which significant information already exists. CA calculate that of the five million compounds on their registry file, only 30% have been cited more than once. At the level of the laymen/non-chemical scientist the number of compounds of interest probably does not exceed half a million.

It might seem that for the layman, the precise unambiguous description of a substance given by a systematic name might be the nomenclature of choice; in practice the complexity of such names mean that they fail to constitute the concise and convenient designation required. In addition for these users, who are often concerned with purchase and the use of purchased materials, the inevitability of trade names complicates the picture.

For the non-chemist, systematic nomenclature is irrelevant. For the non-chemist, trivial names are pronounceable and convenient and will therefore be used as appropriate for communication – a trade name if talking about a purchase, generic names if talking about the active ingredient of a drug or citing in legislation.

For retrieval of information the non-chemist is going to use the name he knows and here lies the biggest danger. For all the reasons described, and for historical ones as well, different information will exist on compounds under different names. The non-chemist is much less likely to know them all; the non-chemist cannot create systematic nomenclature; the non-chemist will almost certainly not be able to relate the equivalence of a compound with two different trivial names through the variety of systematic ones.

What is needed then for the important (non-esoteric) chemicals is for the lay user to be able to identify quickly equivalence between a substance cited by several names, and to locate the possible names for a particular compound. A few directories already exist (e.g. *The Pesticide Manual*) but have two main disadvantages:

- they are not comprehensive, in terms of range of names rather than chemicals covered;
- they are not chemically structured.

It is necessary therefore to develop an international chemical dictionary which provides comprehensive cross-referencing between all names for a single compound and which is structured in such a way as to direct/alert the non-chemist user to closely related molecules. Such a dictionary should be a single dictionary to reduce the number of places to be looked at and to guarantee comprehensivity. The dictionary should include both names and numbers and should be a computerised data base from which relevant subsets can be derived in printed/micro/machine readable forms. The technological and scientific mechanism for developing such a dictionary already exists. WLN and connection tables can be used to collate/identify names for identical materials, e.g. the *Fine Chemicals Directory*, where the WLN is used to bring together all stereo forms of aspartic acid and its metal/inorganic salts. It is therefore essential that the current work to produce registers of important substances, e.g. TOSCA/EINECS/CA registry, is extended to produce a chemical dictionary of this type in which use of a known name will point to the preferred name and a complete set of other terms.

Before developing a chemical dictionary, the question of what constitutes a chemical compound must be carefully considered. As mentioned earlier the conventions adopted by CA in assigning registry numbers exaggerate certain differences and therefore separate information on closely similar structures.

A chemical dictionary should be based on the following hierarchical definition of a molecule: a molecule can be defined in order of complexity as:

(1) a two-dimensional description of a parent compound;
(2) stereo forms, i.e. the three-dimensional molecule;
(3) substitution of atoms in the parent, e.g. as in Na salt, labelled compounds;
(4) additions to the parent, e.g. as in hydrochloride;
(5) proportion of constituent, e.g. mixtures, complexes, polymers.

A chemical dictionary should allow the searcher to start with a name at any of these levels and be cross-referenced to the others with a complete summary of the compounds' use and an indication both of current names and names that have fallen into misuse.

SUMMARY

There is still considerable effort being put into the development of systematic nomenclature rules. Since the systematic nomenclature is of limited use compared with other methods of describing structures this effort would not appear to be necessary. There is also little point in providing new systematic names for compounds which already exist under old ones. Systematic nomenclature should produce a unique name – possibly IUPAC should give way to CA – certainly there is no room for the two. Trivial and semi-systematic nomenclature must continue and it is impossible that standardisation can be achieved for trivial names. Harmonisation of nomenclature systems should be restricted to the maintenance of comprehensive cross-referencing between names to give a complete historical picture of the nomenclature of a compound and its relationships to close derivatives. Acronyms should be discouraged since here lie the greatest dangers for redundancy – so should registry numbers as currently defined.

The chemist has created a Tower of Babel; at least the chemist should document it and take responsibility for it. A chemical dictionary is an urgent need and its development should commence now. It is a difficult but not impossible task and perhaps nomenclature chemists should keep in mind the words of Cardinal Newman, 'A man would do nothing, if he waited until he could do it so well that no one would find fault with what he has done'.

REFERENCES

1. R. Schoenfield, *J. Chem. Inf. Comput. Sci.*, 1980, **20**, 65.
2. R. Walentowski, *J. Chem. Inf. Comput. Sci.*, 1980, **20**, 181.

CHAPTER 17

The use of symbols and nomenclature as an aid to health and safety

R. T. KELLY
Greater London Council, United Kingdom

In much of its work the Scientific Branch of the Greater London Council has to communicate with 'customers' who have little, if any, specialised training in chemistry. To such people, 'layman' as they will be called in this paper, the names of chemicals so familiar and explicable to the chemist are confusing and meaningless, especially when they are used in conversation, as they often must be in the course of the Branch's work. The more complex chemical names are actually intimidating to them and their use impedes effective communication. Even to the chemist, names such as 2-(2,4,5-trichlorophenoxy)ethyl 2, 2-dichloropropionate are clearly too unwieldy for every day use and consequently short names are coined, many of them with only a local currency. Admittedly there are recommended abbreviated forms but these, too, can create difficulties. For example, the pronunciation of 'coumatetralyl', that is 4-hydroxy-3-(1,2,3,4-tetrahydro-1-naphthyl)courmarin, may not be immediately apparent even to a trained chemist. Even the simplest term may be unclear: the difference between methyl and ethyl can be lost in rapid speech and regional variations in pronunciation can obscure the difference between sulphite, sulphate and sulphide.

This paper makes a first attempt at identifying some of the characteristics of chemical naming systems for use by the layman and how a better definition of 'lay' user needs could be obtained and met.

AN ALTERNATIVE TO THE USE OF CHEMICAL NOMENCLATURE

The Scientific Branch of the Greater London Council and the London Fire Brigade have developed a coding scheme which demonstrates that careful consideration of the needs of the ultimate user makes it unnecessary to use chemical nomenclature at all for the communication of health and safety advice.

The scheme was based on about a decade of experience of cataloguing information and is designed to be used by the Brigade at the increasing number of chemical incidents which it is called upon to handle.

The Hazchem code is an initial action code, that is, it does not aim at specifying

(a) what the chemical is; or
(b) what its properties are; or
(c) what methods should have been used to prevent health and safety risks arising; or
(d) what, under ideal circumstances, would be appropriate methods of dealing with a chemical, say in the form of a spillage.

It aims instead at telling the Brigade and other emergency services what are the optimum methods for dealing with a chemical using the materials and techniques which are immediately available. It is based on a combination of letters and numbers whose significance are presented on two sides of a 4 in. × 3 in. card (Figure 1).

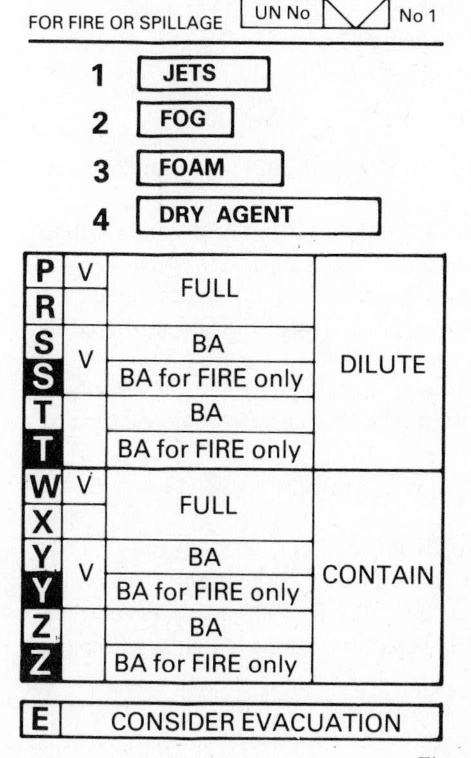

Notes for Guidance

FOG

In the absence of fog equipment a fine spray may be used.

DRY AGENT

Water **must not** be allowed to come into contact with the substance at risk.

V

Can be violently or even explosively reactive.

FULL

Full body protective clothing with BA.

BA

Breathing apparatus plus protective gloves.

DILUTE

May be washed to drain with large quantities of water.

CONTAIN

Prevent, by any means available, spillage from entering drains or water course.

Figure 1

The following examples illustrate its use:

2P

2 means use water spray, fog, foam, dry agent to extinguish the fire, according to the extent of fire, etc.

P means (a) emergency personnel should wear full protection, i.e. chemical protection suits, self-contained breathing apparatus, gloves and boots; (b) the chemical can be safely diluted with copious quantities of water to the draining or sewage system. (c) may react vigorously or violently.

Such a coding would be applied, for example, to sulphuric acid.

3YE

3 means foam or dry agent may be used according to extent of fire.

Y means (a) emergency service personnel should wear self-contained breathing apparatus and protective gloves; (b) the chemical may react violently when involved in a fire or spillage; (c) any spillage should be contained by any means available, e.g. sand, dry earth, etc. and absorbed for disposal.

E means there is a possibility of the hazard spreading beyond the immediate incident area and evacuation of the public should be considered in such areas.

Such a coding would be applied, for example, to ethyl bromide.

4WE

4 means that water must not be used for fire extinguishing; a suitable dry agent only to be applied.

W means (a) emergency service personnel should wear full protective clothing (chemical protection suit, self-contained breathing apparatus, suitable gloves and boots; (b) the chemical may react violently; (c) any spillage should be contained by any means possible, e.g. sand, dry earth etc. and absorbed for disposal.

E means there is a possibility of the hazard spreading beyond the immediate incident area and evacuation of the public should be considered in such areas.

Such a coding would be applied, for example, to acetyl chloride.

The Hazchem code has proved successful in use and has formed the basis for the Hazardous Substances (Labelling of Road Tankers) Regulations 1978 which requires warning signs to be carried when any of some 400 substances are transported. It is also currently being used as a basis for marking buildings within the London area to indicate their contents.

Of course Hazchem, developed for a particular user, covers only a part of a wide spectrum of need for advisory codes on health and safety matters. It is becoming increasingly obvious that improved means are needed for giving advice to the public, specific groups of workers and those involved in environmental protection.

Experience with Hazchem suggests that:

(a) the development of such codes needs to be undertaken with great care so that, for instance, one code does not become confused with another; and
(b) whatever the success of the code, a back-up advisory service may still be necessary.

It has been found that the information given to the Brigade by the Hazchem coding is not in itself sufficiently comprehensive and consequently it has had to be augmented:

(a) In the Brigade operations room detailed information is held on about 17,000 chemicals which are most likely to be encountered.
(b) Appropriately qualified Scientific Branch staff are available round the clock for advice by phone or attendance at incidents.

ANOTHER EXAMPLE OF THE NEED FOR BACK-UP ADVICE

The need to back up the immediate advice given by a marking scheme applies also to more extensive schemes such as The Packaging and Labelling of Dangerous Substances Regulations 1978 which require that containers of greater than a certain size are marked with

(a) the name of the material;
(b) a symbol indicating one of the seven types of risk inherent in the material: Toxic, Corrosive, Harmful, Highly Inflammable, Irritant, Explosive and Oxidising;
(c) an indication of particular risks, there being 58 phrases such as 'Irritating to eyes and skin', 'Causes burns' and 'Reacts violently with water'; and
(d) an indication of the safety precaution required such as 'Keep locked up' and 'Keep container dry'.

Each material covered by the Regulation is listed together with the appropriate symbol, and risk and safety precaution phrases as required.

Despite the comprehensive nature of the regulations the Health and Safety Executive foresaw that they would need to be augmented from various sources.

HOW TO IMPART ADDITIONAL INFORMATION

The task of imparting such information is, in the author's opinion, made more troublesome by the lay user having to employ chemical nomenclature. This complication would be avoided if he had available to him a simpler referencing method. What needs to be developed is a series of 'identifying tags' which can be easily said, heard, read and written. These could cover commercial formulations as well as pure chemicals. They would be designed not as a means of directly imparting information about the chemical or formulation, but as a mechanism:

(a) for facilitating the cataloguing of information;
(b) for marking containers so that enquiries about their contents could be linked to the catalogued information more easily and more certainly than at present; and
(c) for use orally, so that there would be a reduced chance of misunderstanding and misidentification.

POSSIBLE FORMS OF SUCH TAGS

The most obvious form of such tags would be digits and letters, either separately or in combination to produce strings, such as

12345, ABCDE, 12BC98

Initially the author considered the use of such digit or letter tags would be unacceptable only because there appeared a likelihood of confusion with the many systems already in use which employ these forms. Further investigation showed that the use of this sort of 'tag' could be unacceptable for other reasons.

Conrad and Hille[1] discussing the use of long strings of digits as telephone numbers state:

> Low grade mental defectives can cope with only two or three digits, while there are a few eccentrics who can manage twenty or more, usually by rapid encoding of groups of digits into some other form. Within the middle range of ordinary telephone users one can fairly safely say from the considerable research that has been done that half the adult population of this country could not with certainty repeat back a seven-digit number every time, when they had heard it only once.

Hence, there is uncertainty about the general public's ability to repeat accurately strings of digits. The number of digits which an identifying tag would need to have would depend on the number of chemicals and formulations to which it was decided tags would be allocated. If a target of a 100 million materials were chosen, then an eight-digit string would be necessary. There appears to be a strong possibility that in some cases such digit strings would be said incorrectly, an inconvenience when using the telephone but a mistake which could have catastrophic results in emergencies if the string was being used as a means of obtaining advice.

Even if a string of letters and digits is correctly said, there would be a risk that the recipient would hear them incorrectly. Extensive work by Hull[2] on the accuracy of auditory transmission has shown high error rates, unacceptably high in the author's opinion, occurring in the use of digits and letters in such identifying tags. For example, there is confusion between the letters B, C, D, and E and between certain letters and digits, e.g., A and 8. As a means of improving the accuracy of such transmissions, letter-naming techniques (A-Alpha, B-Bravo), as used by the emergency services, has been recommended. However, Hull comments: 'it is clear that these strategies are not employed systematically or automatically by the naive user, but require deliberate learning and practice'. There seems to be a strong *prima facie* case for rejecting identification tags consisting of digits and letters.

The most abvious alternative to digit and letters on such tags is words, themselves producing tags of the form:

APPLE – CITY – DOG

STRUCTURE AND VOCABULARY FOR WORD TAGS

If identifying tags composed of words are to be developed, two questions need to be asked.

(a) What sort of words should be used?
(b) What should be the structure of the tag?

On the first, the fact that the tags are to be used by the general public indicates broadly the type of word which will be acceptable. It has been established that at least two million adults in the country do not have the same word skills as the average nine-year-old pupil in British schools. Some indication of the extent of the literacy problem has been given by Highton[3] who estimates that between 25 per cent and 50 per cent of the British adult population would have some difficulty reading and understanding the following, all of which are quite widely used, in giving advice in health and safety matters:

ignition	explosive
combustible	flammable
by inhalation	harmful
liberates	irritating
toxic	respiratory

The underlined letter combinations 'incombustible' and 'inhalation' are rare and would cause particular difficulty. Thus, the words used for forming the tags need to be drawn as far as possible from commonly used vocabulary.

Two additional requirements probably need to be met by the thesaurus.

(i) Words widely used with a defined meaning in emergency service work, may need to be avoided.
(ii) Words which rhyme, such as 'tide' and 'bride', must be avoided because the tags must be distinct when spoken.

The work of Hull and colleagues[4] at the Medical Research Council Applied Psychology Unit, suggests that:

(a) the number of segments in the tag should not exceed four;
(b) the number of words used to form the segments should be kept as small as possible otherwise increased error in use might result because of confusion between words;
(c) ambiguous word forms should be avoided; this means that not only should, for instance, 'bow' not be used because of two possible pronunciations, but 'bough' should be avoided because of possible confusion when it is contrasted with 'cough', 'though';
(d) the words used should be concrete nouns if possible;
(e) nonsense sentence and semantic relationships within the four segment tags should be avoided;
(f) personal names, which are familiar to most people, would be a good starting segment for such a tag, since their use would differentiate the identifying tag from other codes in common use; and
(g) there would be an advantage on categorising the words used for the other segments in the tag; perhaps the second segment should be chosen from a list containing only animate objects (boy, ant, tiger, etc.) and another segment should be chosen from a list containing only inanimate objects (stone, earth, iron, etc.).

A POSSIBLE ADVANTAGE OF SUCH WORD TAGS

If a system of four word tags could be developed by adding a fourth list of words, say geographical locations (Paris, London, etc.), to the three categories suggested by Hull[2], a significant advantage would accrue.

The significance for identification purposes of four word tags, such as George-Ant-Stone-Paris is independent of the order in which the words are said, provided that no word appears in more than one list. Thus, whether George-Ant-Paris-Stone is said or Paris-Ant-George-Stone, the recipient can rearrange the words into the correct order because he knows that 'Paris' cannot be from list 1 (it is not an animate object) and George cannot be from list 3 (it is not a geographical location).

If each list were to have 200 words consisting of first names (List 1), animate objects (List 2), inanimate objects (List 3) and geographical locations (List 4), identification tags could be fixed to 1600 million individual chemicals or formulations, a number sufficient, in the author's opinion, for present purposes and for growth for some time to come.

INTERNATIONAL IMPLICATIONS

All the above discussion is based on English as the means of communication. Obviously it would be possible for each country to choose its own thesaurus of words best suited to native speakers. How could these separate national lists be linked? Some mechanism for allocating tags would be necessary, at least nationally, to prevent the same tag being used for two different chemicals or to ensure that the same chemical did not have two different tags.

It would be a substantial administrative task, but a feasible one, to make such tag allocations international through a single organisation, using the nationally developed words lists, to ensure for a single material, the appropriate tags in all languages are decided and published at the same time.

FURTHER INVESTIGATIONS

However, decision on these matters must remain unresolved until more fundamental points are settled. Based on the comments the author obtained during the limited consultations he was able to carry out, the three most important are:

(1) Does anything really need to be done for the lay user? There are some who would argue that the problem is not usually as great as the author's experience would suggest.

(2) If something needs to be done, is a specially designed tagging system the best approach? Some may argue that even if such a scheme were practicable it would negate much of what has already been achieved in the field of chemical nomenclature. It may be that the best way forward is to improve the level of knowledge of the mechanisms already available.

(3) If a lay tagging system is needed, are the benefits of trying to operate internationally worth the considerable extra administrative effort which would undoubtedly be involved? Some would argue that we should make haste slowly, allowing each country to develop its own scheme and try for linking later. Others would argue that, with the increased transport of chemicals around the world, if a lay tagging scheme cannot work internationally it is not worth doing at all. At least one consultant saw the solution to the international application already available – Esperanto. He argued that four word lists (perhaps not using all the categories suggested above) based on Esperanto should be drawn up and popularised. Even those who are persuaded that some lay tagging system needs developing may argue that the system outlined above is too cumbersome for everyday use.

All these points need to be considered carefully and efforts made to establish the true needs of the lay user by wide consultation with professional institutions, safety organisations, trade unions, and consumer groups as well as specialists such as phoneticians, coding experts and those knowledgeable in problems of literacy.

A scheme should then be devised which before implementation would have to be extensively tested in the field. In offering the suggested structure for the word tag, Hull[5] emphasised that the work of the Applied Psychology Unit had demonstrated unequivocally that adequate coding schemes cannot be designed without involving the would-be user at all stages to test the validity of any assumption made.

Finally, the professional chemist must be involved in two ways:

(a) to test that any proposed scheme for lay users does not interact adversely with the nomenclature schemes used for trained chemists; and
(b) to make available the considerable expertise already available in devising and administering world-wide nomenclature schemes for chemicals.

If all this were done, it may be that quite a simple mechanism might be developed to improve matters for the lay user. Despite what is said about expert input, it may allow, for example, an adequate numerical code to be developed. This is an attractive proposition because of its easy application across language boundaries. Whatever the outcome it is important, in the author's opinion, that an early start is made in considering the needs of the lay user.

REFERENCES

1. R. Conrad and B. A. Hill, *Post Office Telecommunications Journal,* 1957, **10**, 37.
2. Audrey J. Hull, *Appl. Ergon*., 1976, 7, 75.
3. Private communication.
4. Private communication.
5. Private communication.

CHAPTER 18

Nomenclature: the way ahead

Professor N. LOZAC'H
University of Caen, France

Nomenclature is no longer the only means for the dissemination of chemical information storage and retrieval. Computers may, for example, store structural information through a variety of different coding methods. Nevertheless, the need for accurate chemical names may not disappear within the predictable future and nomenclature methods are still worth study and improvement.

Unfortunately, progress in nomenclature is slow and dependent upon the unpredictable progress of science. It is beyond the capability of any individual, or group of individuals, to predict the nomenclature rules of the future. For a large part, they will depend upon a chemical knowledge which does not yet exist. What is more desirable is a serious reflection on the value of existing rules and on their adaptability to the progress of chemistry.

PRESENT SITUATION

Scientific progress depends upon the concepts and the words used to convey information from one individual to the other. Any scientist appreciates that whatever the value of his personal achievement, he owes an enormous debt to the mass of knowledge acquired previously by mankind. What is perhaps surprising is that although any reasonable person considers that this fact is evident, only a small group concludes that nomenclature still has an active role in the development of science and for this reason the subject deserves much more effort than is generally allowed within the scientific community.

In chemistry it is accepted that the fundamental bases for modern nomenclature are the studies carried out by Lavoisier and his co-workers who dissipated the obscurities of the language used by the alchemists. Insufficient importance was, however, attached to nomenclature and its problems became so crucial that scientific circles were forced to conclude that unless something was done then great difficulties would ensue. The first important collective effort made to improve organic chemical nomenclature resulted in the Geneva Rules. These contained very good ideas but, unfortunately, they did not bring about all the advantages which could reasonably be expected.

This chapter looks forward to changes in chemical nomenclature rather than to the past. Nevertheless a short account of the evolution of nomenclature since the Geneva Congress in 1892 is very instructive as it illustrates both what should be and what should not be done when developing systems. One extremely valuable principle was put forward at Geneva in rule numbered 17. It stated 'numbering of hydrocarbons is maintained in their substitution products'. If this approach had been applied in a consistent manner then nomenclature would now be in a much better state. It is important to appreciate why such a sound principle, which could have been most beneficial, was eventually neglected in favour of more complicated and inconsistent procedures.

Apart from some terminological shortcomings, Geneva nomenclature suffered from two main weaknesses.

(i) Coverage was very incomplete, hardly any guidance was provided for cyclic structures.
(ii) The suggested method of numbering acyclic structures was extremely unwieldy.

Two conclusions can be drawn.

(a) Any general nomenclature system should cover both cyclic and acyclic structures and use similar principles.
(b) A clear and general numbering system is of the utmost importance.

In systematic nomenclature, issues are often obscured because, more or less consciously, two partly contradictory objectives are sought, namely the creation of practical names for currrent communication, and the creation of unique official names for indexing and literature retrieval. These objectives can be reconciled for simple structures in those cases where the official name is simple enough for current communication. On the other hand the complexity of the official name for complicated structures discourages its use for current communication and, at the best, it should be cited at least once in a given paper. A simpler trivial or semi-trivial name could be used in the current text. Semi-trivial names, for instance, are created by associating, according to particular rules, two or more trivial names whose meaning is well known. The semi-trivial name can be considered as an assembly of 'prefabricated' units.

Too many different procedures have been used simultaneously for assembling trivial and/or systematic names, such as addition, substitution, conjunction and fusion. These are the most important types and they are applied in various ways according to the circumstances of use.

Nomenclature is a matter of decision and the present confusing situation has developed from the lack of a clear policy when decisions have been made. Insufficient attention has been paid to the real significance of nomenclature procedures and to a proper choice of the words. An example of this unfortunate situation is the nomenclature used for amines. The same term 'amine' is used both as a functional name (representing NH_3) and as the name of the substituent NH_2. The result of this confusing policy is that methylamine and methanamine are synonyms. This is difficult to reconcile with a logical approach to nomenclature.

Existing rules permit the naming of simple structures and for more complicated structures such as assemblies of rings and chains allow several names. The International Union of Pure and Applied Chemistry (IUPAC) has what can only be described as a non-rule, namely:

> A-61.1 – Hydrocarbons more complex than those envisioned in Rule A-12 composed of cyclic nuclei and aliphatic chains, are named according to one of the methods given below. Choice is made so as to provide the name which is the simplest permissible or the most appropriate for the chemical intent.

A very necessary objective for future nomenclature work is to ensure that well-based rules are developed for the construction of names.

WHAT CAN BE DONE?

Tradition has a powerful influence on nomenclature. Every chemist claims that he is eager to get a general logical nomenclature system. In practice it is with the proviso that the bits of nomenclature he learnt during his early years remains undisturbed. Any attempt to alter significantly the habits acquired in his career is bound to meet with a violent expression of discontent.

Experience indicates that changes in nomenclature should be reduced to a minimum. However, as changes and/or suppressions become necessary it is essential to define the method of working and to examine the real significance of existing nomenclature rules. The objective of systematic nomenclature is to describe a chemical structure by the association of predefined morphemes associated in a precise way. This entails translating the structural formula into a group of words. Various parts of the chemical structure are considered and named in turn, and the corresponding terms are finally assembled in a prescribed order. Thus, at the basis of any nomenclature activity are nomenclature operations. These operations represent chemical reactions which are sometimes real but more often hypothetical such as addition or substitution.

As users of chemical names do not wish to change their habits unless persuaded that they have compelling reasons for doing so, the only practical way for improving nomenclature is to draw as much as possible from existing practice. This could be achieved by taking advantage of existing mental mechanisms and simply omitting those procedures which cannot take place in a general logical scheme.

Some general ideas emerge from this analysis of the present situation and its drawbacks and these can be used as guidelines for the future evolution of nomenclature.

BASIS OF SYSTEMATIC NAMES

Whilst a trivial name is arbitrarily coined in order to designate a chemical substance, often by reference to aspect (e.g. pinacone now pinacol) or its origin (e.g. strychnine), a systematic name is derived from a structural formula. These formulae may give more or less precise information on the real structure and may sometimes derive from arbitrary assumptions. An example is the name cyclohexa-1,3,5-triene which is related to a Kekule formula of benzene which neglects resonance.

Chemical structures, that is to say, atoms linked by bonds, have to be named. The meaning of atoms is clear but the meaning of bonds is far less clear. To be specific it is necessary to restrict the meaning of bond to situations where this so-called feature has a geometrical meaning in a molecular species. A consideration of solid-state inter-atomic forces in metals, ionic crystals and interaction between ions in solution is outside the scope of this chapter. However, a molecular species is not necessarily neutral: it may be ionic, as, for example, the acetate anion.

PRINCIPLE OF HOMOGENEITY

The principle of homogeneity was put forward some time ago by the Chemical Abstracts Service using the phrase 'treat like things alike'. This is a very sound principle but unfortunately it cannot be generally enforced for two reasons.

(i) There may be some discussion about what is 'like' and what is 'not alike'. For instance the name trimethylamine is generally accepted under the assumption that all three methyls should be treated in the same way. From a more general point of view it can be argued that, wherever possible, the basis of any name should be a parent name with a main characteristic group (i.e. the name expressed with an ending). Adopting this point of view leads to the name N,N-dimethyl-methanamine. This method still raises a fundamental question for the user as to whether he bases his name on likeness in the formation of parent names or likeness of treatment of the methyl groups.

(ii) In many cases there is no existing rule which would permit the user to treat like things alike. The following formula clearly illustrates this point:

$$ClCH_2-CH_2-CH\begin{matrix}\diagup CH_2-CH_2Cl \\ \diagdown CH_2-CH_2Cl\end{matrix}$$

1,5-dichloro-3-(2-chloroethyl)-pentane

The three chlorine atoms are in exactly the same situation but the existing rule does not permit the user to treat them in an alike manner.

MODULARITY

The term 'modularity' is related to the idea that a complex object or concept may be entirely divided into independent and well-identified parts. The use of the term independent in this situation is intended to mean that the content of a part is not altered because of the nature of other parts. Advantages of modularity are obvious. It is easier to describe a complex object by cutting it into smaller parts. By analysing a complicated concept then the concept will become easier to understand. Three closely related types of modularity must be dealt with in systematic nomenclature: modularity of rules; modularity of names; modularity of structure.

(i) Modularity of rules

A prerequisite of applying the modularity principle is that every structural feature should have a precise seniority order. The various features, in order of decreasing seniority, should be considered in turn when constructing a name. Any decision, for example, in numbering, made at any given step should be strictly maintained in the course of subsequent nomenclature operations.

The modularity principle is clearly stated in a number of instances, for example in Rule H-3.11: 'Numbering of an isotopically modified compound is *not* changed from that of an isotopically unmodified compound'.

Modularity in the rules is possible for a given knowledge of the structure, stereochemistry and labelling. However, for a given compound, initially described on the basis of incomplete information, the availability of additional information may result in a change in numbering used in the previous name. This can be illustrated by considering a sample of an isotopically substituted 2,3-dichlorobutane which has been described without any knowledge of its stereochemistry, the numbering being decided by its isotopic substitution. If at a later date the stereochemistry of the compound is established then it will have priority for numbering on isotopic substitution according to Rule H-3.11. This could change the number as shown in the following example:

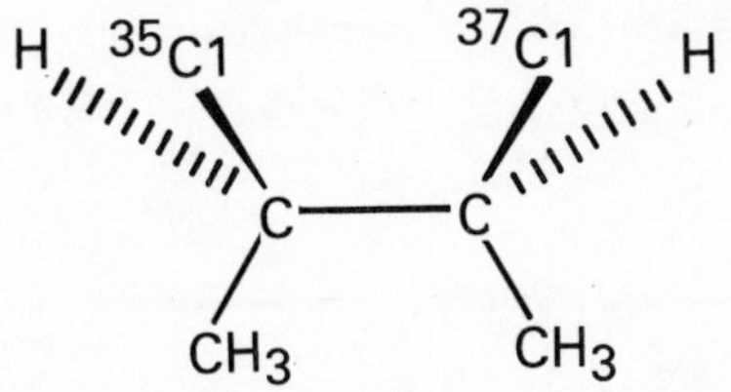

H ^{35}C1 ^{37}C1 H

C—C

CH_3 CH_3

If stereochemistry is ignored, ^{37}C is given the lowest locant according to Rule H-3.22, and the name is then given as:

3-(^{35}Cl), 2-(^{37}Cl) -dichlorobutane

If the stereochemistry of the compound is known, then *R* is preferred to *S* for the lowest locant, and this gives the name:

(2*R*,3*S*)-2-(^{35}Cl), 3-(^{37}Cl) -dichlorobutane

(ii) Modularity in names

It may seem, at first sight, both logical and easy for the modularity of rules to appear clearly in systematic names, any nomenclature operation adding only prefixes or suffixes to the name which comes as a result of earlier operations.

Modularity of names is useful for information storage and retrieval as it permits easy introduction of sub-entries and addition of chemical information in the course of scientific progress. In particular, many systematic names do not give precisely the stereochemistry of a given substance. Modularity in the rules implies that the numbering of the structure and the general form of the name are not altered by the introduction of stereochemical indicators and in practice, this is the case (Rule E-0). Modularity in the name, if strictly applied, would mean that all stereochemical indicators should appear only as prefixes or suffixes to the unchanged non-stereochemical name. This cannot be applied as a general rule. If the stereochemical indicator is related to the parent name then it should be placed before this name, separating it from the substitutive prefixes. No general system of numbering permits the user to distinguish unambiguously atoms in the parent and in the varous substituents.

The situation for isotopically modified compounds is very similar and is clearly exposed in the Rule H-2-22 which reads: 'In a name consisting of two or more words, the isotopic designator may be placed before the appropriate word or part of the word that includes the labelled nuclide(s), unless unambiguous locants are available or are unnecessary'.

(iii) Modularity in structures

The concept that a structure should be cut into smaller parts and to describe those parts of the name is at the very basis of any systematic or semi-systematic nomenclature. The problem which has arisen with the present situation is that too many different linguistic procedures have been used for the same purpose.

The problem is not whether to render systematic names modular but to use modularity in a consistent way. The following three names may be considered as modular:

methyl chloride
chloromethane
naphthaleneacetic acid

The difficulty in the use of these names is that they are built according to three completely different linguistic procedures namely: addition, substitution, and conjunction (a type of double substitution). Any real improvement would require that such differences in terminology are suppressed. For organic molecular non-ionic species, it is possible to use mainly, if not exclusively, substitutive methods. For ionic molecular species, additive (e.g. -ium, ending) or subtractive (e.g. -ate or -ylium endings) methods are often used. The problem here is to limit the number of endings by a proper use of the convention.

The nodal system is an attempt to reduce drastically the number of contradictions, inconsistencies and shortcomings of existing nomenclature. It is not claimed that the system is the best that can be achieved but only that it is necessary to show that such a system can exist and that some hopes can be nurtured for the improvement of the present situation. Details on the system have been published[1,2].

The basic procedures of nodal nomenclature are:

(i) reduction of the structure to a graph of nodes and bonds;
(ii) division of the prepared graph into modules which are either acyclic or cyclic;
(iii) numbering and naming of the modules; and
(iv) assembly of the modules and creation of a complete sequential numbering based on local numbering of all modules in the complete structure.

This operational procedure provides a complete numbering of the structure and a name for the graph. The procedure remains as near as possible to one of the established nomenclature methods in order to facilitate understanding of this new system. The von Baeyer nomenclature of bridged hydrocarbons was considered to be the best possible starting point for this purpose. This does not mean that nodal names are strictly similar to von Baeyer names. The general logic is similar but selection rules and terminology have been amended and extended.

Subsequent stages depend upon the objective set by the user. This may involve:

(i) definition of a general method giving a unique 'official' name for any structure; or
(ii) the definition of rules for assembling trivial and/or systematic names for the description of complicated structures.

DEFINITION OF OFFICIAL NAMES

In the definition of official names, the skeletal atoms together with the hydrogen atoms directly bound to them, constitute the nodes of the graph. After the graph has been named, the nature of the skeletal atoms can be expressed using the well known oxa-aza replacement method. Any atom in the skeleton not cited with an oxa-aza prefix is a carbon atom.

The final stage is to define the nature of the bonds. For double or triple bonds, the usual endings '-ene' and '-yne' can be used without problem. However, the ending '-ene' is also in the current use for fused polycyclic hydrocarbons to indicate the presence of the maximum number of non-cumulative double bonds. This confusion with an indicator of an isolated double bond is not acceptable. Furthermore, for heteroatomic monocycles, the Hantzsch-Widman system prescribes endings which show simultaneously the size of the ring and that the ring contains the maximum number of non-cumulative double bonds.

It is both clearer and simpler, because of the modular nature, to use one specific term to describe one specific concept. The ending '-arene' is one possible choice to state that a structure contains the maximum number of non-cumulative double bonds. The concept can also be used for acyclic systems or for parts of a larger structure using appropriate locants. If it is considered that the ending '-arene' is too specific, because Rule A-12.4 says that it is the generic name of monocyclic and polycyclic hydrocarbons, then another term such as '-axene' may be adopted. The term used is of secondary importance provided that it is unambiguous to the chemist.

If the user thinks it appropriate, then the substitutive method may be applied to parent names constructed according to the nodal method. This is a possibility but not a necessity. If the objective is to construct unique official names, then the use of the substitutive method requires a clear rule which states that parts of the structures are to be described as substituents. The important feature with this system is that the use of the nodal method provides a general and logical procedure which can be used in various ways for the elaboration of nomenclature rules. It belongs to international organisations such as IUPAC or to scientific editors to decide the precise applications of the system.

ASSEMBLY OF TRIVIAL AND/OR SYSTEMATIC NAMES

The procedures outlined in the preceding paragraph will take the analysis of the structure to its extreme. The nature of every atom and of every bond has to be indicated, often by one separate part of the name (term or locant). Often the accumulation of a large number of independent pieces of information may render the name cumbersome and difficult to understand rapidly.

The use of systematic names is considered inconvenient by many users although this problem has existed since the first attempts were made to achieve the unambiguous naming of chemicals. It explains the preference enjoyed by short trivial names. Remembering trivial names has its own limitations and a compromise has to be sought. A practical procedure is to use trivial names for indicating the presence of rather large 'building blocks' such as aromatic rings, monosaccharides or amino-acids. This method may be compared with the use of prefabricated elements in the construction of buildings.

This method is very interesting for specialised nomenclature as it permits the description of very complex structures by relatively short names built from a limited vocabulary of well-known trivial names. However, no general method has yet been devised for treating this type of problem. The nodal system may offer a solution to this problem. An application of the nodal method to the nomenclature of cyclophanes is currently being studied by the IUPAC Commission on Nomenclature of Organic Chemistry. With appropriate conventions, such methods could be extended to polysaccharides and polypeptides.

CONCLUSION

Progress in the development of a system of chemical nomenclature may proceed in a number of different ways but three important prerequisites must be satisfied.

(1) The objective should be defined: official systematic names, or practical semi-systematic names. Simple rules lead to long names whilst short names will need longer rules.
(2) A general method for numbering structures should be defined. Topology comes first, terminology later.
(3) To achieve practical results, as many existing logical methods should be used as possible together with terminology that is not too strange to the general user. It is theoretically possible to start from scratch, for topology and terminology, but such attempts have never met with success in the past.

This course of action may itself create a danger of confusion between existing methods and the new system. The new terminology should retain many of the traditions of the old system to gain easy acceptance, but should be sufficiently different to be recognisable. This is a delicate balance for anyone concerned with nomenclature to maintain.

REFERENCES

1. N. Lozac'h, A. L. Goodson and W. H. Powell, *Angew. Chem.*, 1979, **91**, 951.
2. N. Lozac'h, A. L. Goodson and W. H. Powell, *Angew. Chem., Int. Ed. Eng.*, 1979, **18**, 887.

Appendix 1. Further Reading

Compiled by Dr G. P. MOSS
Queen Mary College, London, United Kingdom

IUPAC AND IUB RECOMMENDATIONS

Nomenclature of Organic Chemistry, Sections A, B, C, D, E, F and H, 1979 edn (The Blue Book), Pergamon Press, 1979. Section A: Hydrocarbons; Section B: Heterocyclic; Section C: Characteristic Groups (C, H, O, N, F, Cl, Br, I, S, Se, Te); Section D: Other Elements (Metals, P, As, Sb, Bi, Si, B); Section E: Stereochemistry; Section F: Natural Products; Section H: Isotopically Modified Compounds. The previous edition (1971) had only Sections A, B and C.

Biochemical Nomenclature and Related Documents, (The Compendium), Biochemical Society, 1978.

Enzyme Nomenclature, 1978, Academic Press, 1979.

Nomenclature of Inorganic Chemistry, 2nd edn, (The Red Book), Butterworths, 1971.

How to Name an Inorganic Substance, Pergamon Press, 1977.

Compendium of Analytical Nomenclature, (The Orange Book), Pergamon Press, 1978.

Manual of Symbols and Terminology for Physicochemical Quantities and Units, Pergamon Press, 1979.

OTHER IUPAC AND IUB RECOMMENDATIONS NOT INCLUDED IN THE ABOVE (see *Chem. Internat.* 1982, (5), 15, for fuller index)

Note. All IUPAC recommendations are now published in *Pure Appl. Chem.*, and IUB recommendations in *Eur. J. Biochem.* Many IUB recommendations have also been published in other biochemical journals. An asterisk indicates a provisional recommendation.

Revision of the extended Hantzsch-Widman system of nomenclature for heteromonocycles, *Pure Appl. Chem.*, in press.

*Designation of non-standard classical valence bonding in organic nomenclature, *Pure Appl. Chem.*, 1982, **54**, 217.

*Glossary of terms used in physical organic chemistry, *Pure Appl. Chem.*, 1979, **51**, 1725.

*Nomenclature for straightforward (organic) transformations, *Pure Appl. Chem.*, 1981, **53**, 305.

*Nomenclature of tetrapyrroles, *Pure Appl. Chem.*, 1979, **51**, 2251; *Eur. J. Biochem.*, 1980, **108**, 1.

*Nomenclature for five- and six-membered ring forms of monosaccharides and their derivatives, *Eur. J. Biochem.*, 1980, **111**, 295; *Pure Appl. Chem.*, 1981, **53**, 1901.

*Nomenclature of unsaturated monosaccharides, *Eur. J. Biochem.*, 1981, **119**, 1; 1982, **125**, 1; *Pure Appl. Chem.*, 1982, **54**, 207.

*Nomenclature of branched-chain monosaccharides, *Eur. J. Biochem.*, 1981, **119**, 5; 1982, **125**, 1; *Pure Appl. Chem.*, 1982, **54**, 211.

Nomenclature of tocopherols and related compounds, *Eur. J. Biochem.*, 1982, **123**, 473; *Pure Appl. Chem.*, 1982, **54**, 1507.

*Abbreviated terminology of oligosaccharide chains, *Eur. J. Biochem.*, 1982, **126**, 433; *Pure Appl. Chem.*, 1982, **54**, 1517.

*Polysaccharide nomenclature, *Eur. J. Biochem.*, 1982, **126**, 439; *Pure Appl. Chem.*, 1982, **54**, 1523.

*Nomenclature of vitamin D, *Eur. J. Biochem.*, 1982, **124**, 223; *Pure Appl. Chem.*, 1982, **54**, 1511.

*Nomenclature of retinoids, *Eur. J. Biochem.*, 1982, **129**, 1.

*Symbols for specifying the conformation of polysaccharide chains, *Eur. J. Biochem.*, in press.

*Abbreviations and symbols for specifying the conformation of polynucleotide chains, *Eur. J. Biochem.*, in press.

Corrections and Additions to *Enzyme Nomenclature 1978, Eur. J. Biochem.*, 1980, **104**, 1; 1981, **116**, 423; 1982, **125**, 1.

Units of enzyme activity, *Eur. J. Biochem.*, 1979, **97**, 319; 1980, **104**, 1.

Atomic weights of the elements 1979, *Pure Appl. Chem.*, 1980, **52**, 2351.

Nomenclature of isotopically modified inorganic compounds. *Pure Appl. Chem.*, 1981, **53**, 1887.

Nomenclature of hydrides of nitrogen and derived cations, anions and ligands, *Pure Appl. Chem.*, 1982, **54**, 2545.

Naming of elements of atomic number greater than one hundred, *Pure Appl. Chem.*, 1979, **51**, 381.

Nomenclature of inorganic boron compounds, *Pure Appl. Chem.*, 1972, **30**, 681.

Presentation of NMR data, A: Proton Spectra, *Pure Appl. Chem.*, 1972, **29**, 625.

Presentation of NMR data, B: Spectra from Nuclei other than Protons, *Pure Appl. Chem.*, 1976, **45**, 217.

Symbolism and nomenclature for mass spectroscopy, *Pure Appl. Chem.*, 1978, **50**, 65.

Definitions and symbols of molecular force constants, *Pure Appl. Chem.*, 1978, **50**, 1707.

Use of abbreviations in the chemical literature, *Pure Appl. Chem.*, 1980, **52**, 2229.

Nomenclature dealing with steric regularity in high polymers, *Pure Appl. Chem.*, 1966, **12**, 643.

Basic definitions of terms relating to polymers, *Pure Appl. Chem.*, 1974, **40**, 477.

Nomenclature of regular single-strand organic polymers, *Pure Appl. Chem.*, 1976, **48**, 373.

Stereochemical definitions and notations relating to polymers, *Pure Appl. Chem.*, 1981, **53**, 733.

*Nomenclature for regular single-strand and quasi single-strand inorganic and co-ordination polymers, *Pure Appl. Chem.*, 1981, **53**, 2283.

CHEMICAL ABSTRACTS

Index Guide 1982, Appendix IV, paras. 101–293, Chemical Abstracts Service.

Parent Compound Handbook 1976 and Supplements, Chemical Abstracts Service.

J. E. Blackwood, C. L. Gladys, A. E. Petrarca, W. H. Powell and J. E. Rush, Unique and unambiguous specification of stereoisomerism about a double bond in nomenclature, *J. Chem. Doc.*, 1968, **8**, 30.

J. E. Blackwood and P. M. Giles, Jr, Chemical Abstracts stereochemical nomenclature of organic substances in the Ninth Collective Period (1972–1976), *J. Chem. Inf. Comput. Sci.*, 1975, **15**, 67.

N. Donaldson, W. H. Powell, R. J. Rowlett, Jr, R. W. White and K. V. Yorka, Chemical Abstracts Index Names for chemical substances in the Ninth Collective Period (1972–76), *J. Chem. Doc.*, 1974, **14**, 3.

M. F. Brown, B. R. Cook and T. E. Sloan, Stereochemical notation in coordination chemistry: Mononuclear complexes, *Inorg. Chem.*, 1975, **14**, 1273.

M. F. Brown, B. R. Cook and T. E. Sloan, Stereochemical notation in coordination chemistry: Mononuclear complexes of coordination number seven, eight and nine, *Inorg. Chem.*, 1978, **17**, 1563.

BRITISH STANDARDS INSTITUTION

BS 1831:1969, *Recommended Common Names for Pesticides.*

BS 2474:1965, *Recommended Names for Chemicals used in Industry.*

BS 3502:1978, *Schedule of Common Names and Abbreviations for Plastics and Rubbers.*

BS 4589:1970, *Abbreviations for Rubber and Plastics Compounding Materials.*

BEILSTEIN

Stereochemical conventions, p. xxi.
Beilstein Dictionary (German-English), p. xiv.

AGROCHEMICAL NOMENCLATURE

The Pesticides Manual: A World Compendium, 6th edn (Ed. C. R. Worthing), British Crop Protection Council, 1979.

Pesticide Index, 5th edn (Ed. W. J. Wiswesser), Entomological Society of America, 1976.

Index Phytosanitaire, 16th edn, (Ed. R. Bailly and G. Dubois), Association de Coordination Technique Agricole, 1980.

Weed Control Handbook, 6th edn, Vol. 1, (Ed. J. D. Fryer and R. J. Makepeace), Blackwell Scientific Publications, Oxford, 1977.

Agricultural Chemicals Book, 4 vols. (Ed. W. T. Thomson), Thomson Publications, Fresno, California, 1977-9.

PHARMACEUTICAL NOMENCLATURE

USAN and the USP Dictionary of Drug Names 1981, US Pharmacopoeial Convention Inc., 1981.

International Non-proprietary Names (INN) for Pharmaceutical Substances, Cumulative List No. 5, World Health Organization, 1977.

Approved Names 1977 and Supplements, British Pharmacopoeia Commission, HMSO, London.

Index Nominum 1980, Swiss Pharmaceutical Society, 1980.

Organic-chemical Drugs and their Synonyms (Ed. M. Negwer), Akademie-Verlag, Berlin, 1978.

Non-proprietary Names for Pharmaceutical Substances, Technical Report Series No. 581, World Health Organization, 1975.

WISWESSER LINE NOTATION AND COMPUTER HANDLING TECHNIQUES

The Wiswesser Line-Formula Chemical Notation (WLN), 3rd edn (Ed. E. G. Smith and P. A. Baker), Chemical Information Management Inc., 1975.

G. Palmer, Wiswesser Line-Formula Notation, *Chem. Br.*, 1970, **6**, 422.

J. E. Ash and E. Hyde, *Chemical Information Systems,* Ellis Horwood, Chichester, 1975.

M. F. Lynch, J. M. Harrison, W. G. Town and J. E. Ash, *Computer Handling of Chemical Structure Information,* McDonald, London, 1971.

HISTORY OF NOMENCLATURE

M. P. Crosland, *Historical Studies in the Language of Chemistry*, 1962; Heinemann, 1962; reprinted, Dover, 1979.

W. D. Flood, *The Origins of Chemical Names*, Oldbourne, 1964.

P. E. Verkade, Etudes historiques sur la nomenclature de la chimie organique: Parts I to XVI, *Bull. Soc. Chim. Fr.*, 1979, 215 II and previous references.

BOOKS

R. S. Cahn amd O. C. Dermer, *Introduction to Chemical Nomenclature*, 5th edn, Butterworths, 1979.

Nomenclature of Organic Compounds: Principles and Practice (Ed. J. H. Fletcher, O. C. Dermer and R. B. Fox), Advances in Chemistry Series, No. 126, American Chemical Society, 1974.

J. E. Banks, *Naming Organic Compounds: A Programmed Introduction*, 2nd edn, Saunders, 1976.

J. G. Traynham, *Organic Nomenclature: A Programmed Introduction*, Prentice-Hall, 1966.

D. Hellwinkel, *Die Systematische Nomenklature der Organischen Chemie: eine Gebrauchsanweisung*, Springer Verlag, Berlin, 1974.

N. Lozac'h, *La Nomenclature en Chemie Organique*, Masson et Cie, Paris, 1967.

M. L. MacGlashan, *Physicochemical Quantities and Units: the Grammar and Spelling of Physical Chemistry*, 2nd edn, Royal Institute of Chemistry, 1971.

Quantities, Units and Symbols, Report of the Symbols Committee of the Royal Society, 1975.

MISCELLANEOUS

The Merck Index, 9th edn, Merck and Co. Inc., 1976.

R. S. Cahn, (Sir) Christopher Ingold and V. Prelog, Specification of molecular chirality, *Angew. Chem.*, 1966, **78**, 413; *Angew. Chem., Int. Ed. Engl.*, 1966, **5**, 385 and 511.

V. Prelog and G. Helmchen, Basic principles of the CIP-system and proposals for a revision, *Angew. Chem.*, 1982, **94**, 614; *Angew. Chem., Int. Ed. Engl.*, 1982, **21**, 567.

N. Lozach'h, A. L. Goodson and W. H. Powell, Nodal nomenclature: General principles, *Angew. Chem.*, 1979, **91**, 951; *Angew. Chem., Int. Ed. Engl.*, 1979, **18**, 887.

Index